GENOME ANALYSIS AND BIOINFORMATICS

GENOME ANALYSIS AND BIOINFORMATICS
A Practical Approach

T.R. Sharma
Principal Scientist (Biotechnology)
National Research Centre on Plant Biotechnology
IARI Campus, Pusa, New Delhi-110012, India

TECHSAR™
NEW DELHI

Published and Printed by

TechSar Pvt. Ltd.
4435-36/7, Ansari Road, Daryaganj,
New Delhi – 110 002 (India)
E-mail: tech_info@techsarworld.com
Website: www.techsarworld.com

ISBN 978-93-80026-36-7

Reprint 2025

Dedicated
to
my mother
and
my wife

Foreword

The research in biological sciences is undergoing tremendous change because of the availability of numerous genomics tools. Major impetus in genome research came with the decoding of human genome for which draft sequence was published in June 2000. Simultaneously, genomes of many microbes, mammals, fungi and plant species are being sequenced in different laboratories of the world. Now more than 4000 organisms genomes have been decoded and sequences are available in public domain. To make use of this massive sequence information, and discover the hidden knowlege, extensive analysis of this sequence inforamation, and discover the hidden knowledge, extensive analysis of this sequence information is highly desirable.

Post-genomics era will witness complete transformation, the research is being conducted in different species. Bioinformatics tools would be helpful in locating DNA sequences in the GenBank simply from their accession numbers, making alignments of two or more than two sequences, performing similarity searches for unknown sequences in the GenBank, assembling short sequence reads and developing consensus sequences, finding genes and markers *in sillico* and in performing comparative genome analysis.

The book entitled, **Genome Analysis and Bioinformatics – A Practical Approach** has been written in such a way so that the teachers, scientists and student, both in computer and biology should understand the basic principles of sequences analysis and bioinformatics. In this era of information technology and biotechnology, sharing and dissemination of information has relatively become easy. The author, Dr. T.R. Sharma has put some complex issues related to genomics and bioinformatics in a very simple and understandable language. I am sure it will go a long way in making an impact on biological research and teaching in the country.

(MANGALA RAI)
Seceratary, Department of Agricultural Research & Education and
Director General, Indian Council of Agricultural Research
Ministry of Agriculture, Krishi Bhavan, New Delhi- 110001

Preface

With the decoding of complete genome sequence of many organisms like human, mouse, rice *Arabidopsis* and many microorganisms new vistas of research have started not only in these species but also in related species by comparative and functional genomics approaches. The published data on genome sequence of these organisms reveals a comprehensive picture of the DNA codes and positions and distribution of all the genes, repetitive sequences and centromeres in the genome. Determination of individual DNA components in the form of ATGC base pairs by using various sequencing techniques and parallel advances in recombinant DNA technology helped in the study of an organism at whole genome level. DNA sequence information is becoming an indispensable tool in modern biology. However, efficient use of this information can only be performed by understanding basics of genomics and bioinformatics. For the application of genome sequence information for the benefits of human being, trained human resource is required.

Being associated with the sequencing of rice genome under International Rice Genome Sequencing Project from 2000 to 2005, I was supposed to learn basics of high throughput genome sequencing, sequence assembly, annotation and performing comparative genome analysis at the National Research Centre on Plant Biotechnology (NRCPB), IARI, New Delhi and at the Cold Spring Harbor Laboratory, NY, USA as part of my Training on High throughput DNA Sequencing and bioinformatics. It was indeed a great challenge to understand and then apply genomics and bioinformatics techniques in the rice genome analysis. It was also an opportunity to read extensively many good books on molecular biology and bioinformatics. Subsequently, I got an opportunity to design and teach a course on Bioinformatics for the Post Graduate School of the Indian Agricultural Research Institute, New Delhi. During these years, I was using various books and web resources as reading and teaching material. Though I found various interesting and very good books on genomics, molecular biology and bioinformatics, many times it was difficult to understand basics of important topics and then explaining those to the students.

I was always looking for a book which includes basic topics on genomics and bioinformatics in a very simple communicable language so that any beginner can start these subjects without any difficulty. To achieve this objective a hybrid book on "Genome Analysis and Bioinformatics': A Practical Approach" was planned. It will definitely act as beginner guide to both biologists who are interested in bioinformatics and to the computer experts who want to make career in bioinformatics. An effort has been made to explain various aspects in a simplest possible manner using well labeled diagrams and figures. For the application of bioinformatics tools, wherever necessary, practical exercises have been included along with different steps so that one can use basic minimum bioinformatics tools without any difficulty. It would be a basic book for researchers and students of B.Tech/M.Tech (Bioinformatics & Biotechnology), all post-graduate students of biology, genetics, genomics & biotechnology and even for MCAs etc.

At this moment, I would like to thank various individuals who inspired and encouraged me at one or other occasions during my academic life. First of all, I am really very thankful to my teacher, Prof. B.M. Singh, Former Dean, College of Agriculture, Agricultural University, Palampur from whom I have learnt to put my thoughts in a very simple manner. I am grateful to Dr. N.K. Singh, Principal Scientist, NRCPB, with whom I am involved in sequencing rice, tomato and Pigeonpea genomes and shared very good research experiences. I am also very much thankful to the former and present Project Director of NRCPB, for their whole-hearted support and also providing me very congenial teaching and research environment.

For the preparation of this book many Research Associates and Research Fellows working in my different projects have helped me in the initial phase of collection of reading material. I thank all of them for their help. However, I am specially thankful to Mr. D.K. Gupta and Ms H. Sonah for collecting material on chapters related to sequence alignment, phylogeny analysis and DNA marker analysis, respectively. I also thank Mr. Jitender Pareek for the redrawing of various figures included in this book. I am also grateful to many anonymous individuals who have provided very good reading material on the World Wide Web and also to various genome centres that are making genome sequences available in the public domain.

I would like to place on record my special thanks to my wife Mrs. Madhu Sharma, daughter Asuda and son Akshaj for their full co-operation and endurance during these past few years since, I could not spare much time for them because of my pressing academic commitments. I would also like to thank my parents and brothers who were the main source of inspiration and support for me to reach upto this level.

I wish that this book should reach to all parts of the country so that students, researchers and teachers can make best use of it and give their feedback for its further improvement.

T.R. Sharma

Contents

Abbreviations

AFLP	Amplified Fragment Length Polymorphism
ANOVA	Analysis of Variance
AS-PCR	Allele-Specific Polymerase Chain Reaction
ATP	Adenosine Triphosphate
BAC	Bacterial Artificial Chromosome
BLAST	Basic Local Alignment Search Tool
BSA	Bulked Segregant Analysis
CAPS	Cleaved Amplified Polymorphic Sequences
cDNA	Complementary DNA
cM	centi-Morgan
DAF	DNA Amplification Fingerprint
DNA	Deoxyribose Nucleic Acids
ELISA	Enzyme-Linked Immuno Sorbent Assay
EST	Expressed Sequenced Tag
GMO	Genetically Modified Organism
GO	Gene Ontology
GUS	Glucuronidase
InDel	Insertion Deletion
ISSR	Inter-Simple Sequence Repeat
ITS	Internal Transcribed Spacer
JSC	Jaccard's Similarity Coefficient
NCBI	National Centre for Biotechnology Information
ORF	Open Reading Frame
PCR	Polymerase Chain Reaction

PAC	Pi-Artificial Chromosome
QTL	Quantitative Trait Loci
RAPD	Random Amplified Polymorphic DNA
RFLP	Restriction Fragment Length Polymorphism
RGA	Resistant Gene Analogue
RIL	Recombinant Inbred Line
RNA	Ribose Nucleic Acids
RT-PCR	Reverse Transcription PCR
SCAR	Sequence Characterized Amplified Region
SNP	Single Nucleotide Polymorphism
SSR	Simple Sequence Repeat
STS	Sequence-Tagged Site
TIGR	The Institute of Genomic Research
UPGMA	Unweighted Pair Group Method with Arithmetic Mean
VNTR	Variable Number Tandem Repeats
WGS	Whole-Genome Shotgun
YAC	Yeast Artificial Chromosome

1 Introduction

Rapid advances in genome research in the recent past have resulted in generation of large set of data for DNA and protein sequences from different prokaryotic and eukaryotic genomes. The entire set of chromosomal genetic material of an organism is known as genome. World over, as of now (October, 2008), more than 4000 ongoing genome projects are at different stages of sequencing. Around 48% of these projects are on bacterial genome sequencing followed by the eukaryotic genome sequencing (Fig.1). Massive information is being generated in terms of genome sequences in different organisms which will help in understanding basic and applied research in biology. Molecular biology research actually started with the discovery of double helical structure of DNA by Watson and Crick in 1953. By that time, it was also known that the DNA is made up of four nitrogenous bases known as adenine (A), guanine (G), thymine (T), and cytosine (C). The complete sequences of A, T, G and C, in the genome can be determined by using automated DNA sequencing machines. Since, DNA is a very long molecule, no machine can sequence a DNA molecule completely in one go. On an average, a machine can generate a sequence of ~ 500 base pairs in a single sequencing reaction. There are three major applications of perfoming sequence analysis of an organism.

i) Identification of mutation or SNPs by sequencing of short regions of DNA.

ii) Obtaining sequence of full length gene along with its upstream and downstream regulatory regions.

iii) Sequencing of complete genomes of the organism.

Automation of Sanger's dideoxy method of DNA sequencing in the last decade, laid the foundation of genome research in different organisms. The genomes of a large number of microorganisms and other species including human, have already been sequenced. Determination of individual DNA

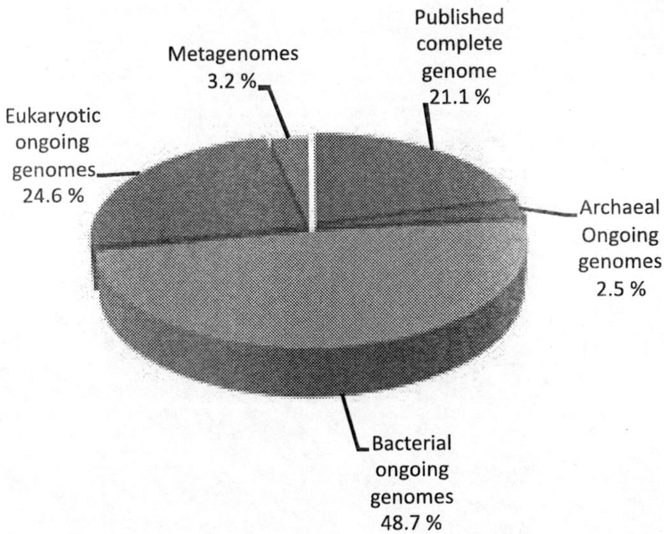

Fig.1. Different genome projects in progress world over as on October, 2008 (source www http://www.genomesonline.org/gold.cgi).

components in the form of ATGC bases by using various sequencing techniques and parallel advances in recombinant DNA technology helped in the study of an organism at whole genome level. DNA sequence information is becoming an indispensable tool in modern biology.

The huge wealth of information in the form of DNA and protein sequences and publications on molecular biology is stored in the data banks. Major public data banks which take care of the DNA and protein sequences are GenBank in USA (http://www.ncbi.nlm.nih.gov), EMBL (European Molecular Biology Laboratory) in Europe (http://www.ebi.ac.uk/embl/) and DDBJ (DNA Data Bank) in Japan (http://www.ddbj.nig.ac.jp). These public databases are continuously growing. This rapid growth in DNA sequencing data is due to the fact that many collaborative international programmes have started during the past few years to sequence complete genomes of various organisms. The whole genomes of many microorganisms have already been sequenced by the Institute of Genome Research (TIGR) which can be seen on their website www.tigr.org.

With the advent of various advanced methodologies, sequencing complete genomes of organisms have become reality which has generated huge amount of sequence data in the public domain. Therefore, analysis of complete genome of different organisms can be undertaken by using genomics and bioinformatics tools.

Bioinformatics is the computational analysis of biological data, consisting of the information stored in the form of DNA and protein sequences in various biological databases. The National Centre for Biotechnology Information (NCBI 2001) has explained the term bioinformatics as follows.

"Bioinformatics is the field of science in which biology, computer science and information technology merge into a single discipline. In bioinformatics, three important areas are generally considered. These are:

(i) development of new algorithms for the assessment of relationships among members of large data sets;

(ii) analysis and interpretation of various types of nucleotides, amino acids and protein domains and

(iii) efficient access and management of various types of information by implementation of different tools. Bioinformatics is thus defined as the computational systems which are used for the collection, storage and analysis of biological information. These include software systems that take in DNA sequence data, database systems that store data, and software systems that analyze stored data (Sobral, 1997).

The history of computational biology (bioinformatics) goes back to the 1920s when scientists were already thinking of establishing biological laws solely from data analysis (Lotka, 1925). However, the development of powerful computers, and the availability of experimental data (for example, DNA or amino acid sequences and three-dimensional structures of proteins), that can be readily analyzed by computation, launched bioinformatics as an independent field. Today, practical applications of bioinformatics are readily available through the World Wide Web (www), which are widely used in biological research. As the field is rapidly evolving, the very definition of bioinformatics is still the matter of some debate. The relationship between computer science and biology is a natural one due to the following reasons:

i. The phenomenal rate of biological data being produced, provides challenges of its storage, analysis, and making it accessible to one and all.

ii. The nature of data often requires statistical and computational methods. This applies in particular to the information on the building protein models and of the temporal and spatial organization of their expression in the cell encoded by the DNA.

Analyses in bioinformatics focus on three types of datasets: genome sequences, macromolecular structures and functional genomics experiments (e.g. microarray data). However, bioinformatics tools are also applied to various other data for getting meaningful results.

Bioinformatics tools are now easily available to the biologists with the advent of internet and various web browsers on world wide web. These tools are indispensable for any genome sequencing centre. For the time being, most of these softwares are available free of charge to the public Institutions. However, once a software becomes popular in scientific community, its developer starts putting restrictions for its limited use or makes it available on payment basis. Beside these software, small scripts written in Perl and Java languages are also used to help biologists in handling large genome sequences at various stages of data generation, assembly and annotation. Apart from sequence analysis software, Laboratory Information Management System (LIMS) is also required in big genome centres to keep track of the clones and sequences, because of the involvement of various complex procedures, steps and large number of individuals working on different aspects.

Once a high quality sequence is obtained one has to ask an important question whether this is a new sequence or the sequence similar to other DNA sequences available in the databases. For getting an answer to this question, one has to perform database search for sequence comparison. All sequence searching methods rely on the basic concepts of alignment and distance between the sequences and pair-wise sequence alignment is performed. This type of sequence comparison is generally performed with BLAST (Basic Local Alignment Search Tools), which compares unknown sequence against all the sequences available in the database (http://www.ncbi.nlm.nih.gov/).

Once similarity search is performed between unknown sequence and the database sequence to find per cent homology between them, it is obvious to know how these sequences are related to each other. The sequences derived from two closely related organisms show more similarity at DNA level compared to the sequences derived from the distantly related organisms. To find an evolutionary relationship among sequences derived from different organisms, a phylogenetic tree is constructed. Such evolutionary tree can also be constructed on the basis of phenotypic markers, molecular markers or on the basis of DNA /protein sequence information. A typical phylogentic tree comprises nodes, branches and termini of the branches. For constructing a phylogenetic tree the PILEUP option of GCG package is more commonly used. Besides, DNA STAR software (www.dnastar.com) also has options to construct the tree from different DNA or protein sequences. However, web based tools like MacClade (//www.phylogeny.arizona.edu/macclade/) can also be used for evolutionary studies of different organisms based on their DNA sequences. Similarly, bioinformatics tools can be used for protein function analysis by database search. Mining SSR markers and SNP markers from the EST or genome sequences is also one of the important areas of research in computational biology.

Simply determining four alphabets (ATGC) of DNA sequences of any organism has no value until some meaning is derived from this by finding number of genes present within the sequence using various gene prediction software. Gene prediction is a complex work and there is no algorithm which can exactly predict the true exons in a DNA sequence. Basically two major criteria are taken into consideration while predicting a gene. 1) identification of structural elements such as start/ stop codon and splice sites of the unknown sequence and 2) performing homology search against protein, EST and cDNA database to identify potential coding regions. For gene prediction, many software are available on the world wide web. However, very commonly used software is GENSCAN which has been developed by MIT, USA (http://www.genes.mit.edu/GENSCAN.html). This software is freely available on Web for online analysis of DNA sequences. The output obtained from the GENSCAN is then used for gene annotation by using BLAST search on the public or private DNA sequence databases to find out the matches to the unknown query sequence with millions of sequences available in the GenBank.

Development of suitable algorithms is an important part of bioinformatics. The techniques and algorithms have been specifically developed for the analysis of biological data, for instance, the dynamic programming algorithm for sequence alignment is one of the most popular programmes among the biologists. The sequence information generated worldwide is stored systematically in different types of databases. Hence, it is also important to understand the database management systems and database architecture.

Considering the information explosion in genomics and bioinformatics this book is not an exhaustive one. For that matter, no book can be exhaustive on a subject like genomics and bioinformatics where the information is growing in an exponential manner. However, it is a modest attempt to explain some of the basic topics of the genome analysis using genomics and bioinformatics techniques to both biologists and computer experts in simplest possible manner.

References

Lotka A.J. (1925), Elements of physical biology. Williams and Wilkins, Dover Publications Inc.

Sobral, B.W.S. (1997), Common language of bioinformatics. Nature. 389:418.

Watson, J. D. and F.H.C. Crick (1953), Molecular structure of nucleic acids. Nature. 171:737-738.

2

High Throughput Genome Sequencing

Generation of DNA sequences is a routine method now. However, about 30 years back, several efforts were made to develop sequencing technology. The very first attempt on DNA sequencing is that of Wu and Taylor, who sequenced 12 bases from the cohesive ends of phage l DNA in 1971. Then, an improved method was developed by Sanger *et al.* (1977) i.e., 'plus and minus' method used for sequencing 5386 bps of phase jX174 genome. Subsequently, they developed the most famous chain termination or dideoxy method of DNA sequencing (Sanger *et al.,* 1977). In the same year, chemical method of DNA sequencing was developed which works on the use of chemical reagents for making base specific cleavage of DNA by Maxam and Gilbert (1977). These methods have been briefly explained in the following sections.

DIDEOXY METHOD OF DNA SEQUENCING

This method is also known as Sanger's dideoxy method of DNA sequencing. The unknown DNA fragment which is to be sequenced is called template. The template is first made into single standard form before sequencing. The basic sequencing reaction is performed in four test tubes which consist of various components besides the templates. These component are i) a small stretch of DNA sequence called primers, ii) DNA polymerase enzyme, iii) a mixture of four deoxynucleotide triphosphate (A,T,G,C), iv) one of the dideoxynucleotide i.e. either ddATP, ddTTP, ddGTP or ddCTP labeled with radioactive substances (S^{35}) or non-radioactive substances (Dig or biotin). At appropriate temperature condition, the DNA polymerase enzyme extends the primer sequences by adding a deoxynucleotide (base) one after another complementary to the base present in the template. The synthesis of new DNA strand continues until a dideoxynecleotide (ddNTPs) is added in the complementary DNA strand. The addition of dideoxy nucleotide stops further elongation of complementary strand because of lack of OH group at its 3′ end (Fig. 1). During sequencing reaction, the chain termination of newly synthesized strands result in the generation of different sized DNA fragments, ending with labeled ddNTPs. After reaction is complete the reaction mixture of all the four tubes (specific to each base A,T,G,C) are loaded adjacent to each other on a poly acrylamide sequencing gel. The four lanes specific to ddATP, ddCTP, ddGTP and ddTTP produce fragments of varying length upon electrophoresis and autoradiography. The position of bands in the gel is used to directly read DNA sequences from bottom to top as shown in Fig. 2.

The automated DNA sequencing method is based on Sanger's dideoxy method with little variation and known as "cycle sequencing". In this method all the ddNTPs are labeled with different colour fluorescent dyes. These dyes are TAMRA-ddATP for A (gives green fluorescence), R6G-ddTTP for 'T'(gives red fluorescence), R110-ddGTP for G (gives black or yellow fluorescence) and Rox-ddCTP for 'C'(gives magenta fluorescence). Because of this, all four reactions can be run in a single tube and can be separated in a single lane of the gel. The DNA fragments are detected at the

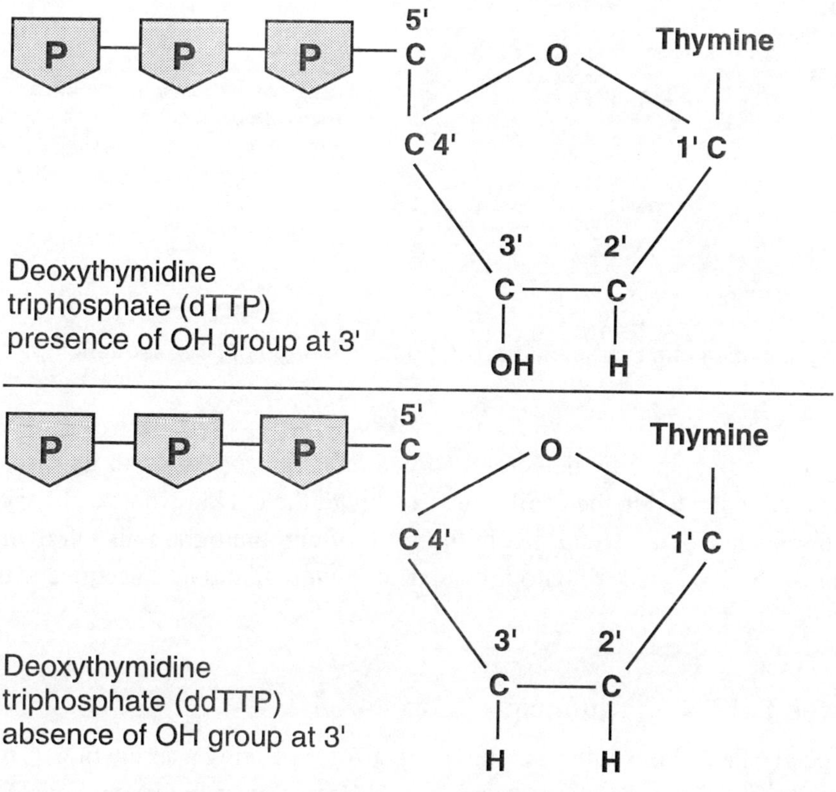

Fig. 1. Difference in the structure of dNTPs and ddNTPs which helps in chain termination during sequencing reaction

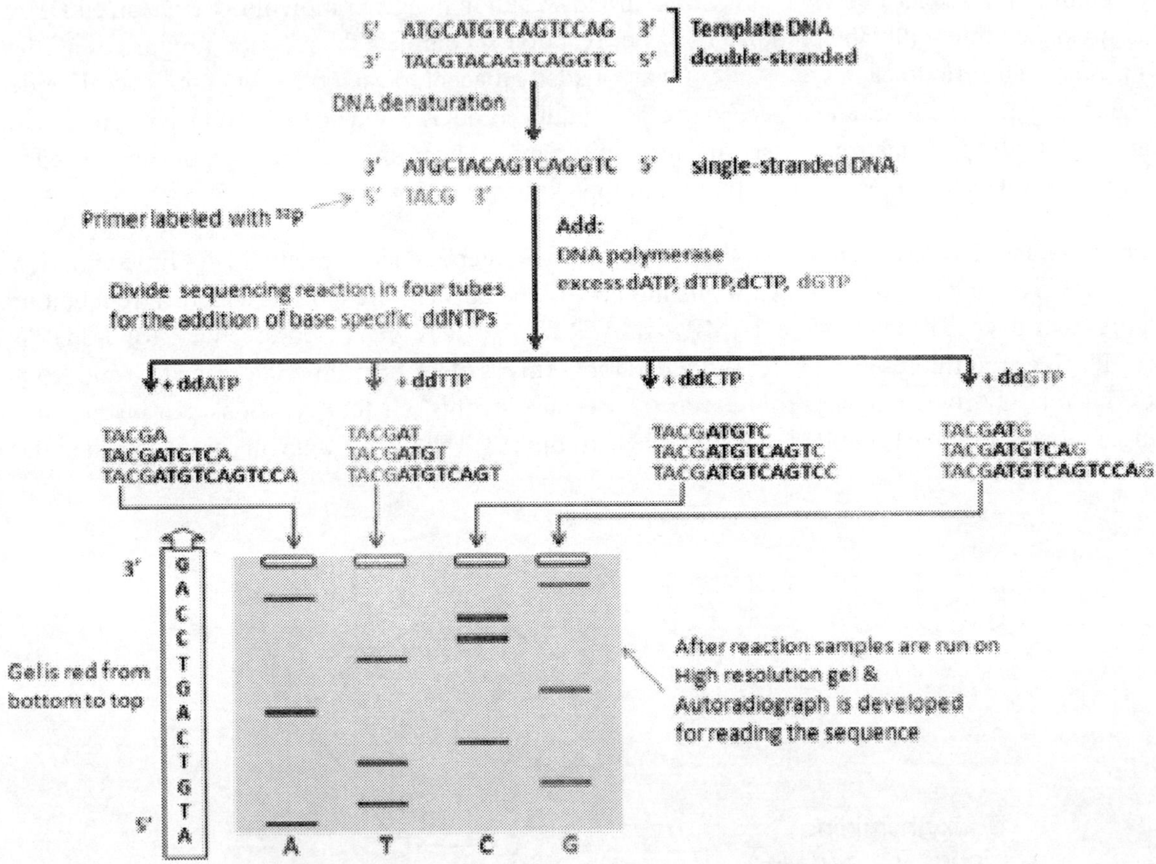

Fig. 2. Schematic representation of dideoxy method of DNA sequencing.

bottom of the gel by using specific detectors. Now a days, instead of 'slab gel', the sequencing reaction products are separated in the capillaries filled with the gel known as capillary based DNA sequencers. The sequence data is recorded in the form of chromatogram also known as Standard Chromatogram File (SCF). This chromatogram data is converted into DNA sequence form by using computational algorithms.

Chemical method of DNA sequencing

Chemical method or Maxam and Gilbert method of DNA sequencing uses chemicals to break DNA molecules at specific bases, thus creating fragments of different sizes. In this method, DNA molecule to be sequenced is radio-labeled at 5′ -PO_4 position by using phophatase and ATPs. The sequencing reaction is devised into four tubes along with a fifth reference tube (Table 1).

Table 1. The contents of different tubes used for sequencing reaction

Tube No.	Chemical added in each tube	Reaction
1.	Dimethyl sulphate	alters methylates guanine at the N 7 position
2.	Acid	alters either adenine or guanine
3.	Hydrazine	alters either thymine or cytosine
4.	Hydrazine along with NaCl	alters cytosine base pair
5.	NaOH only	Reference

For removing altered base pairs from the sequencing reaction, piperidine is added in each tube. Piperidine also breaks the DNA molecules at the sugar residue from the point of altered nucleotide thus making different sized fragments of the DNA. The mixture of DNA fragments are separated on high resolution sequencing gels by loading the contents of all the four tubes in adjacent lanes. The gels are basically made from the polyacrylamide and urea (also called denaturation gels) which helps in fractionation of the fragments based on their size. After electrophoresis, the gels are exposed to X-ray film for developing autoradiographs of the DNA bands from which sequence is read (Fig.3).

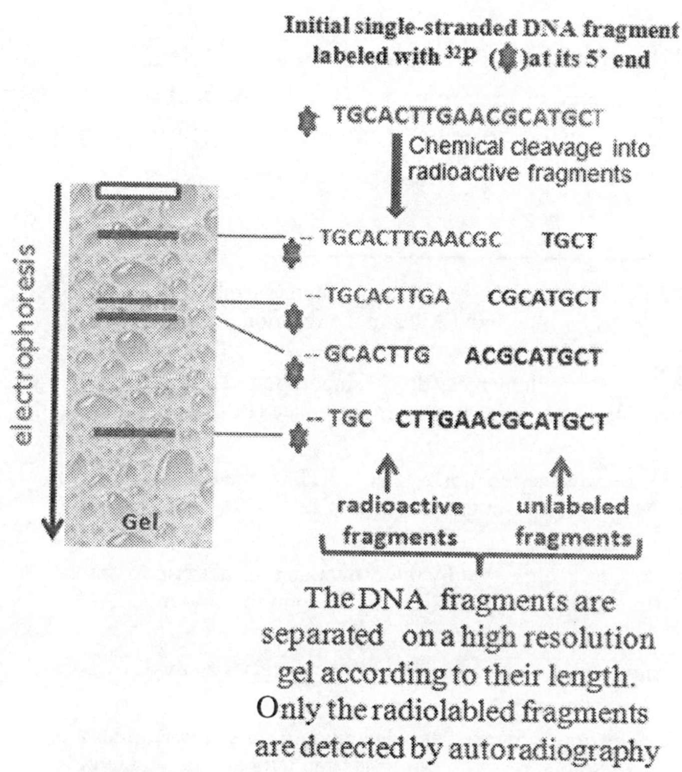

Fig. 3. Basic steps of Maxam-Gilbert DNA chemical sequencing method.

Pyrosequencing

This is another method of DNA sequencing which is now becoming popular for high throughput genome sequencing. The pyrosequencing method has been developed by Mostafa Ronaghi and Pal Nyren (1990) and is based on the sequencing by synthesis principle. In this method one strand of the DNA acts as a template and its complementary strands are synthesized enzymatically by detecting the activity of DNA polymerase enzymes by using another chemiluminescent enzyme. Using this method, addition of one base pair at each step is detected in the newly synthesized DNA strand. In this method the template DNA is immobilized on a solid support (Streptavidin coated magnetic beads) and after the reaction the solutions of A, G, T and C are sequentially added and removed. During this process, when the first unpaired base of the template is complemented by the nucleotide solutions, it produces fluorescence. Therefore, the sequence of the template is determined based on the chemiluminescence signals produced by the sequencing solutions. Various components of pyrosequencing reactions include: i) single stranded DNA template hybridized to a sequencing primer, ii) enzymes, DNA polymerase, ATP sulfurylase, luciferase and apyrase, iii) the adenosine 5′ phosphosulfate (APS) and luciferin. The results are obtained in the following manner as explained in the flow digram (Fig. 4).

The main limitation of this method is generation of short read length as compared to the dideoxy method. It is difficult to sequence more complex GC rich region of the DNA using this method. It is commonly used for re-sequencing of the genome and sequencing prokaryotic genomes where genome size is small and also less repetitive.

Addition of one of the four deoxynuleotide triphosphates (dNTPs) and ATPs in the reaction

↓

Corrected complementary dNTPs incorporated on the template by DNA polymerase release of pyrophosphate (PPi) stoichiometrically

↓

Qualitative conversion of PPi to ATP by the ATP sulfurylase in the presence of luciferase

↓

Generation of visible light by the conversion of luciferin to oxyluciferin (the amount of light is in proportion to the amount of ATP)

↓

Detection of light using specific cameras and its analysis by software

↓

The enzyme apyrase degrade unincorporated nucleotides and ATPs so that reaction can be started with another nucleotide

Fig. 4. Different steps used in pyrosequencing

HIGH THROUGHPUT GENOME SEQUENCING

High throughput genome sequencing strategies are used for sequencing whole genome of the organisms. Two basic methods used to sequence large genomes are explained below:

1. Whole Genome Shotgun Method

The Whole genome shotgun (WGS) method is one of the most efficient methods mainly used in sequencing prokaryotic genomes. The genomes which are less repetitive in nature can be easily sequenced with WGS method and assembled by using computational tools. In this method, high quality genomic DNA is isolated from the target organism. The DNA is then randomly fragmented in 2kb and 10 kb sized DNA fragments to construct 2kb and 10kb shot gun libraries in suitable sequencing vectors. Large numbers of shotgun clones are randomly sequenced by using automated DNA sequencing machines. The sequenced data is assembled by using computational methods. Different steps used in WGS approach are given in Fig.5. The major advantage of this method is that it does not require construction of genetic and physical maps of the genome, which is a time

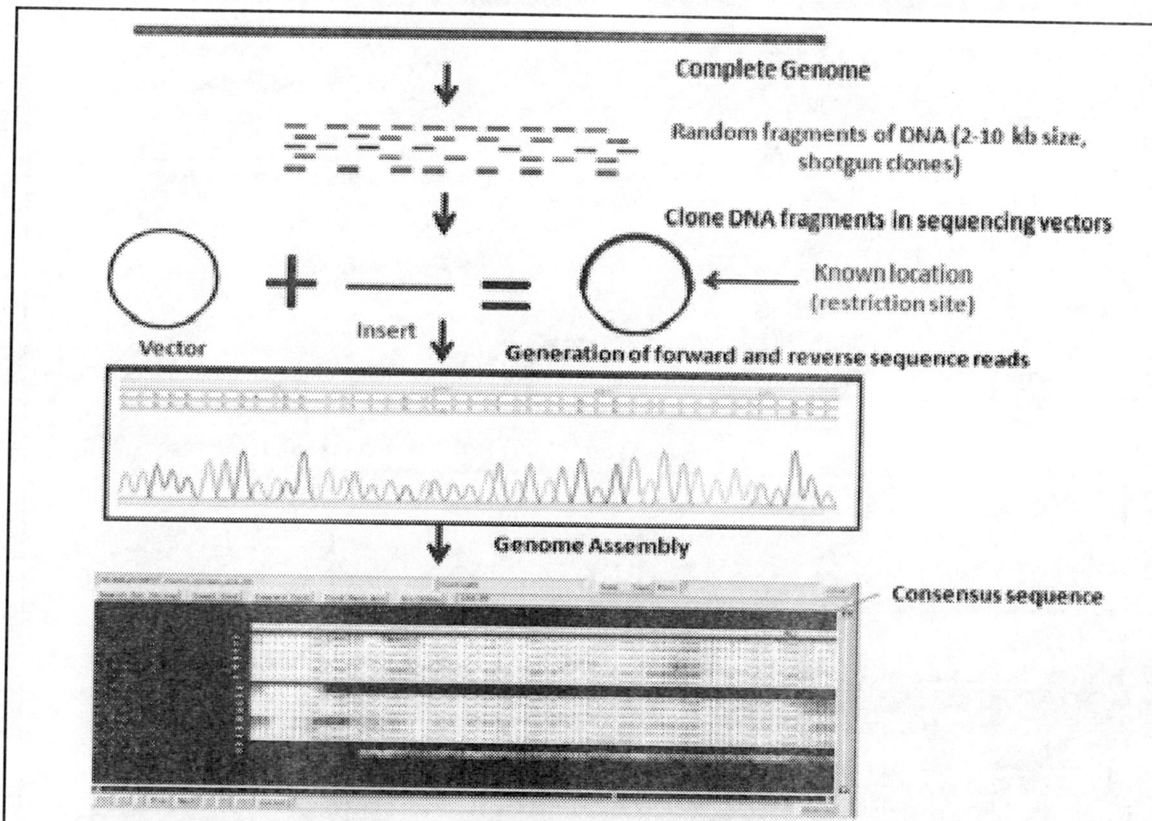

Fig. 5. Different steps used in whole genome shotgun sequence approach

consuming and laborious process. However, the main disadvantage of this approach is that the assembly of sequence reads in the form of chromosome specific pseudomolecules and resolving repeats are difficult in more complex genomes.

2. Hierarchical Sequencing Method

Hierarchical Sequencing Method is also known as clone-by-clone approach or systematic approach of sequencing large genomes (Fig. 6). In this method, construction of genetic linkage map of an organism is a pre–requisite. So the availability of genomic resources in terms of DNA markers and BAC/YAC libraries are very important to construct maps of the plants. Once linkage map is available in plants, it is used for the construction of physical map with the help of large insert BAC/YAC libraries. The BAC clones from a specific region of the physical maps are used for the preparation of shotgun clones as explained briefly in the following sections.

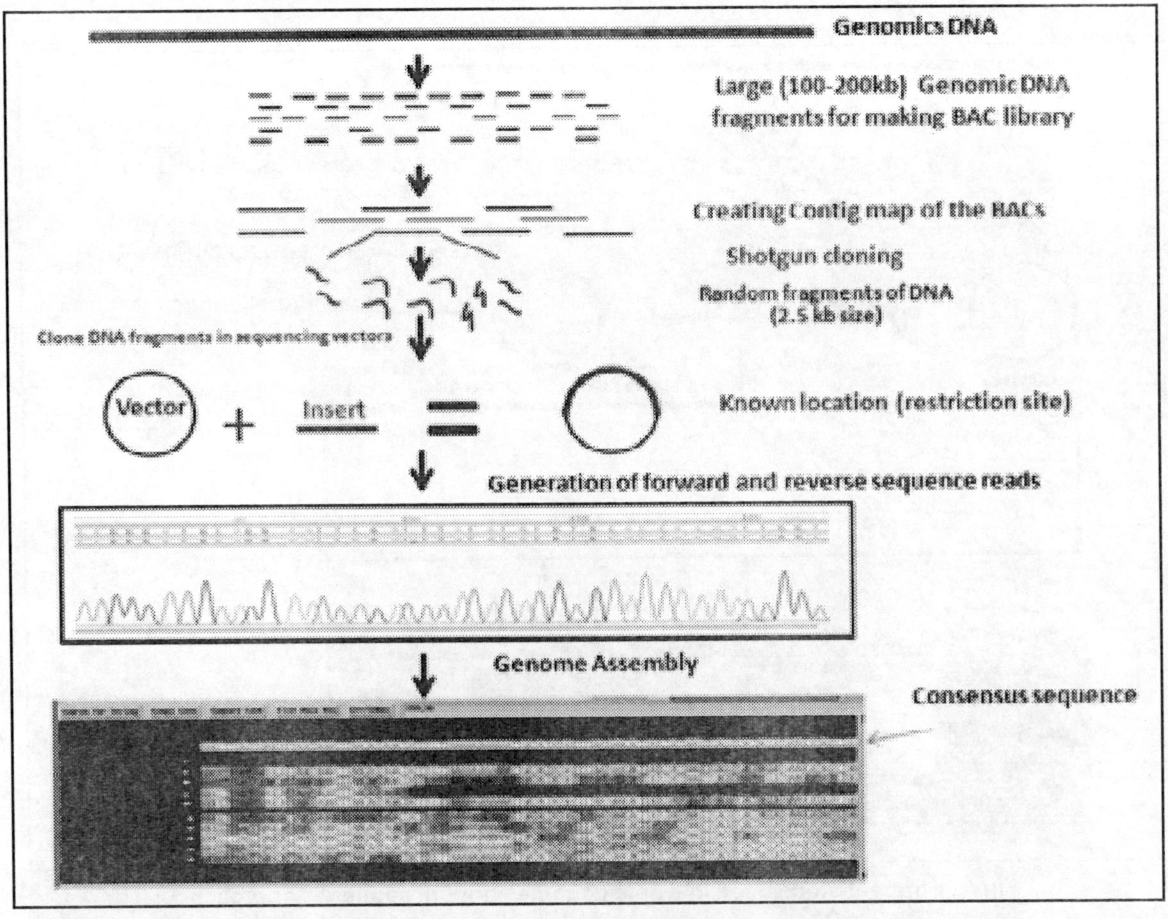

Fig. 6. Different steps used in clone-by-clone approach of genome sequencing.

i) Construction of Physical Maps

Physical mapping means cloning entire genome in large insert vector and then arranging all the cloned fragments in terms of their chromosomal positions by using genetic markers as a probe then constructing minimum path of the large insert clones. The basic requirements for the construction of a physical map are:

- Availability of highly saturated genetic and molecular maps.
- Require large insert cloning vectors.
- High throughput methods of clone picking, culturing and selection.
- Powerful software to construct minimum tiling paths (MTPs).

Different types of large insert vectors used for the construction of physical maps are Lambda phage (insert size 20-30kb), cosmids (35-45kb), BACs and PACs (bacterial and PI artificial chromosome, respectively) able to contain 100-300 kb insert size and YAC (yeast artificial chromosomes) which can contain an insert of 200-1,000 kb. However, YAC vectors are not preferred because these are often chimeric (contain 2 DNA fragments), unstable because of internal deletions and are difficult to purify. Different number of BAC and BIBAC clones in agricultural crops like *Indica* rice (21,078), *Japonica* rice (23,040), *Arabidopsis* (10,368), Soybean (85,944) and Cotton (>200,000) have been used for constructing physical maps. Once physical maps are constructed, next step is to make shotgun or sub clone libraries for sequencing individual BAC clones.

ii) Shotgun Cloning

High quality BAC DNA is isolated and used for making shot gun libraries. Different steps used for isolating BAC DNA are given in Fig. 7. The random DNA fragments ranging from 2 to 5 Kb sizes are obtained by physically breaking of BAC DNA using different methods like sonication, nebulization or hydro-shearing followed by gel separation.

Methods of Generating Random DNA Fragments

A. Sonication

In this method high quality DNA is placed in a buffer in an eppendorf tube or microcentrifuge tubes. The tube is placed into ice-cold water bath for keeping it cool during sonication. A cup-horn sonicator can be used for DNA fragmentations for a varying number of ten-second bursts at continuous power and maximum output. Since temperature increases during sonication which may result in varying sizes of DNA fragments, the temperature of the ice-cold water bath should be monitored and maintained for getting good results.

Centrifuge bacterial culture containing BAC of desired size insert

↓

Add saline solution to the pellet & vortex

↓

Centrifuge and discard supernatant

↓

Alkaline lysis of bacterial cells

↓

Plasmid DNA isolation

↓

Quantification of DNA on Agarose gel

↓

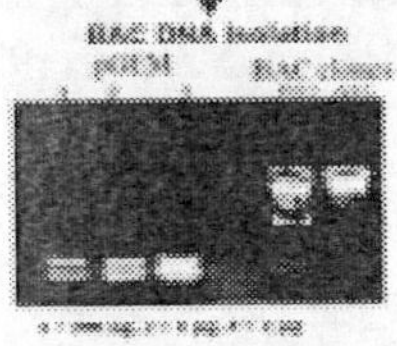

Fig. 7. Different steps used in high quality BAC DNA isolation for use in shotgun cloning.

B. Nebulization

High quality buffered DNA solution (20-50 μg) along with 20-50% glycerol is placed in a nebulizer (Fig.8) which is kept in an ice cold water bath. The inert nitrogen gas is passed through the DNA solution at a pressure of 6 -10 psi for 2-3 minutes. The gas pressure and time has to be standardized empirically for either cosmid or plasmid DNA fragmentation. However, 6.5 psi for 2-3 minutes gives 2-5 kb sized fragments.

C. Hydroshearing

Hydroshearing is the process of passing high quality buffered DNA solution through a small needle of a syringe repeatedly. This leads to physical breakage of the DNA molecules into small fragments. DNA fragments in the size range of 1.0 to 2.5kb are generated in the DNA solution and separated using gel elution technique.

Size-selection of Fragmented DNA Using Electrophoresis

Whole contents of fragmented DNA either by nebulization/sonication/hydroshearing are separated by agarose gel electrophoresis. The sheared DNA gives a smear in the gel lanes and the lanes are buldged out at a specific point where maximum fragments of desired DNA size are accumulated as

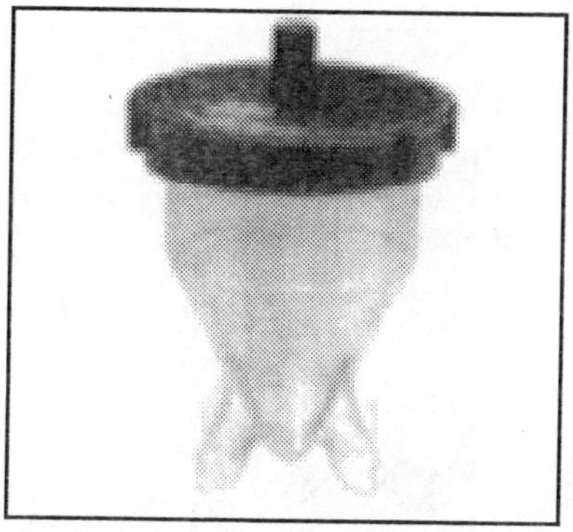

Fig. 8. Nebulizer used for random fragmentation of genomic DNA.

shown in the (Fig. 9, lane 2). Gel slices containing desired size DNA band is cut with clean blade under UV light. DNA fragments are separated from the gel by melting it in warm water, ethanol precipitated and again loaded on a gel to determine the exact size (Fig. 9, lane 3). The purified desired sized fragments are used for end repairing and cloning in a suitable vector.

DNA Ligation

DNA fragments are ligated with appropriate linearized vector like pBluescript or pUC by incubating in the presence of rATP and T4 DNA ligase. The sonicated or nebulized fragments are ligated to the Bluescript or pUC vector by incubation at 4°C overnight for random shotgun cloning. The proper insert to vector ratio are determined for the ligation reaction.

Transformation

Transformation of ligated products is performed by using either electro - competent *E. coli* cells (DH10B, Invitrogen) using Gene Pulser (Bio Rad) or by using chemical method of transformaton. The transformed *E. coli* cells arc plated on LB medium for blue/white screening. White colonies are picked up for one 96 well plate from each BAC clones and DNA is isolated and quantified.These templates are then used for sequencing to check for the possible contamination of bacterial chromosomal DNA or vector DNA. If the extent of contamination is less in the library, it is used for large scale production of shotgun clones for high throughput sequencing. Various criteria used for determining the quality of a shotgun library are given below.

Criteria for Quality Check of a Shotgun Library

Sequencing one 96 well plate

- Identification of contaminants after BLAST search for
  - Any match with *E. coli* genomic DNA

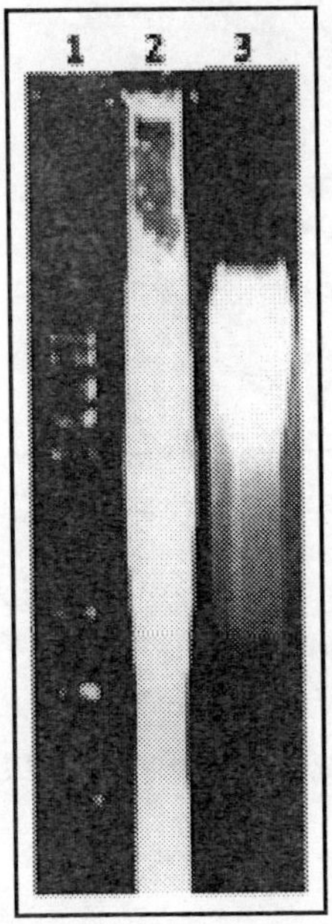

Fig. 9. BAC DNA sheared with nebulization (Lane 2) and size selection (Lane3) to get 2-5 Kb size fragments for shotgun library. Lane 1, DNA size marker.

– Any match with pBeloBAC11 (BAC vector DNA)

– 100% match with pUC19 or any other sub cloning vector

– Significant match (E-value < e^{-20})

- Average mean read length
- Percentage of success

Library passed for large scale shotgun clone production: If overall contamination is < 10%, mean read length 500 bp and percentage of success is more than 80%.

Template Preparation for Sequencing

Before the isolation of plasmid DNA for sequencing, archive of each clone are maintained as glycerol mounts. For this purpose the culture boxes are removed from the incubator shaker and 50 µl bacterial culture is aspirated by using multichannel pipette and added to the 96 deep well plates containing 40% glycerol. The cultures are vortexed, spun down briefly and stored at –80° C freezer. Remaining bacterial culture is centrifuged at 2700 rpm to collect the pellets. One ml saline solution (0.15M NaCl) is added to each well by using multichannel pipette and vortexed vigorously to suspend the bacterial pellets. The samples in the boxes are centrifuged at 2700 rpm for 5 min and supernatant is discarded. Alkaline lyses of bacterial cells are obtained by standard protocol and DNA is isolated by using magnetic beads DNA isolation protocol (or any other standard protocol) and stored at 4° C until use. Quantification of template DNA (5 samples randomly selected from each 96 well plate) is done on 0.8% agarose gel for setting up a sequencing reaction. A summary of different steps used for shotgun library preparation and preparation of template for sequencing are given in Fig. 10.

iii) Automated DNA Sequencing

Once templates are ready, sequencing is performed by the standard protocols of cycle sequencing using big dye terminator chemistry and other sequencing chemistry. DNA templates are first reacted with Terminator Ready Reaction Mix (ABI) in a total volume of 10 ml as per manufacturer's instructions. The plates are kept in PCR machine using standard reaction (PCR) conditions. After 35 reaction cycles, the extension products are purified by ethanol precipitation method and air-dried. Post-PCR clean-up is necessary to remove unutilized dye terminators which otherwise affect quality of the sequence. The DNA pellets are dissolved in 20 µl double distilled sterilized H_2O before loading on automated DNA sequencing machine. The sequence data are collected and analyzed. Different steps used in cycle sequencing are summarized in Fig. 11. In a single sequencing reaction, on an average, 500 bp sequence read is obtained in the form of a standard chromatogram file. A

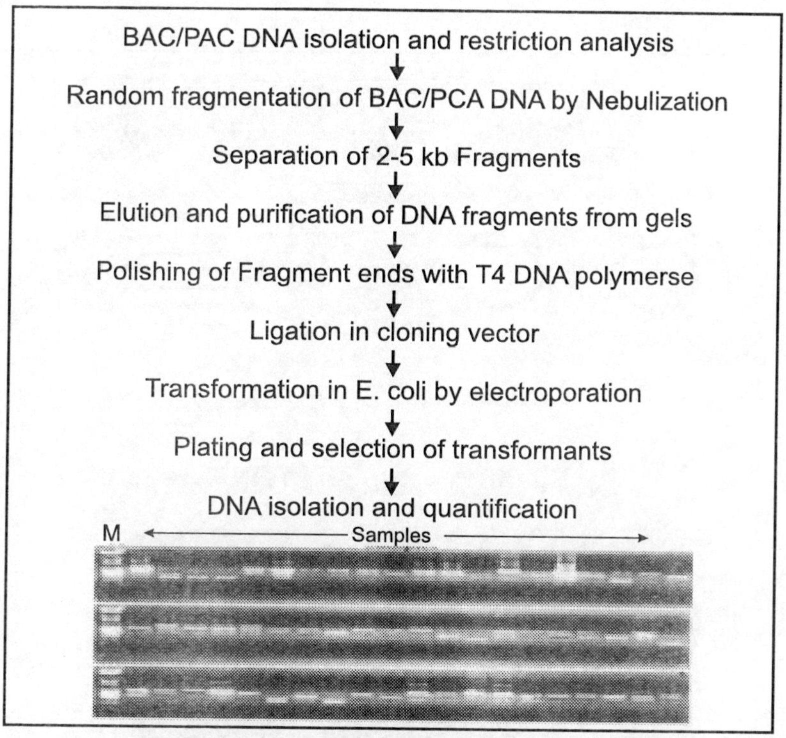

Fig. 10. Steps used for the preparation of shotgun libraries and preparation of DNA template for sequencing.

Table 2. Genome sizes and predicted genes in different organisms.

Organism	Size (bases)	Predicted Genes (No)
Escherichia coli (Bacterium)	4.6 Million	3,000
Sacchromyces cerevisiae (Yeast)	15 Million	6241
Neurospora crassa (Fungus)	39.9 Million	10,000
C. elegans (Nematode)	100 Million	20,621
C. briggsae (Nematode)	104 Million	19,507
Drosophila melanogaster (Fruit fly)	120 Million	13,647
Anapheles gambiae (Mosquito)	280 Million	13,600
Strongylocentrotus purpuratus (sea urchin)	814 Million	23,300
Ciona intestinalis (Ascidian sea squirt)	160 Million	15,800
Fugu rubripes (Pufferfish)	365 Million	38,000
Monodelphus domestica	~3.5 Billion	~20,000
Mouse	3 Billion	30,000
Human	3 Billion	31,000
Arabidopsis thaliana	100 Million	25,498
Rice	389 Million	35,544

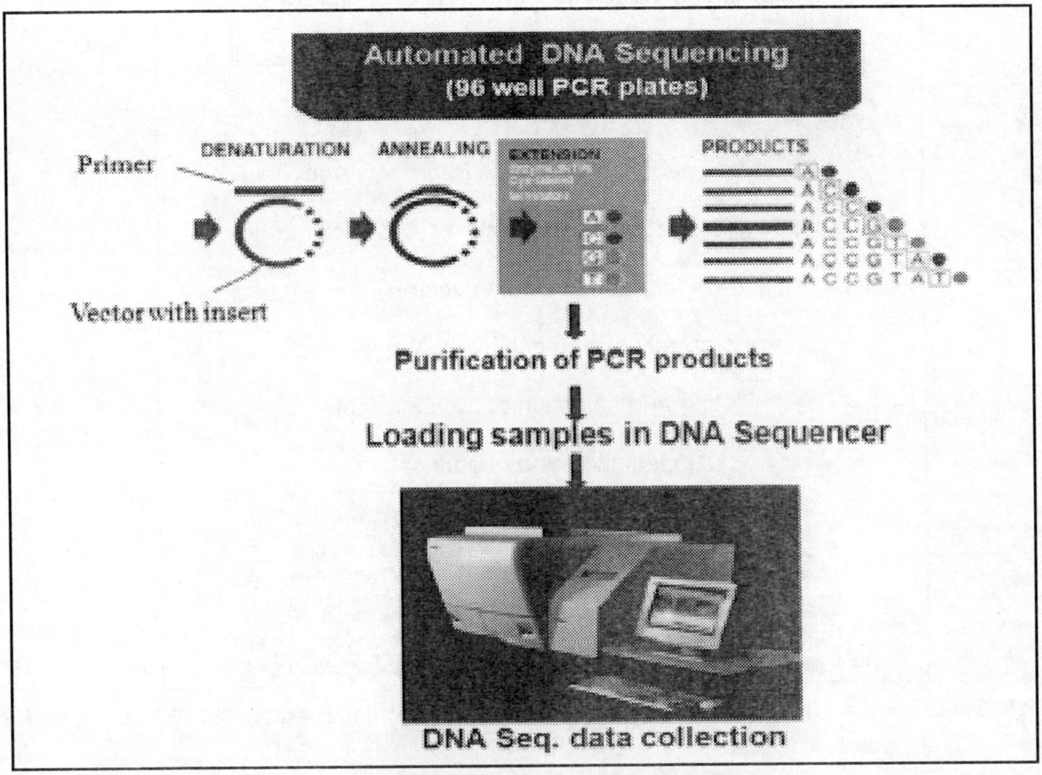

Fig. 11. Steps used in cycle sequencing.

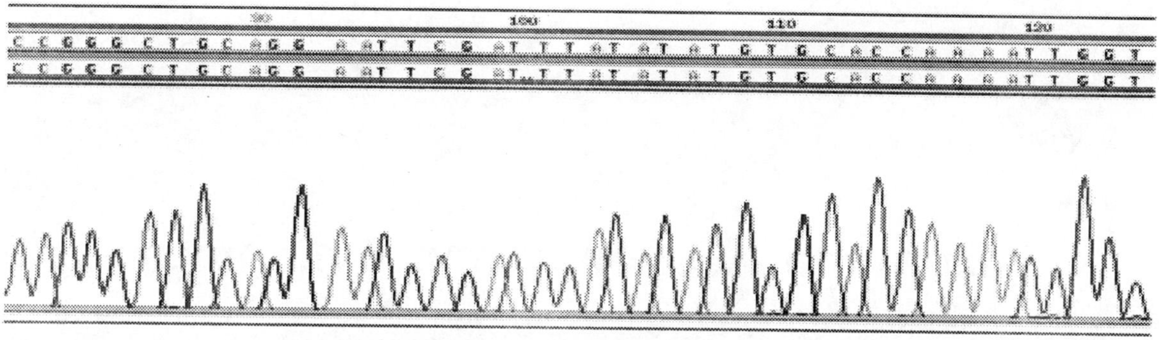

Fig. 12. A typical trace file view of DNA sequences.

typical view of the DNA sequence chromatogram is given Fig. 12. Once sufficient sequence data are generated for each BAC clone, it is assembled with various computational tools and further polished by using different genome finishing methods before submission to the GenBank. Many genomes have already been sequenced by using this technique. Some of the organisms, their genome size and number of genes predicted in their sequences are given in table 2. This data is being used in functional and comparative genome analysis in different molecular biology laboratories of the world.

Suggested Reading

Brown, S.M. (2000), Bioinformatic: A Biologist's Guide to Biocomputing and the Internet. Eton Publishing, Natick. MA , USA.

Lewin B. (2006), Genens IX. Jones and Bartlet Publishers, Inc Sudbury, Massachusetts.

Maxam A.M. and W. Gilbert (1977), A new method for sequencing DNA. PNAS.**74**: 560-564.

Nelson D.L. and M.M. Cox (2008). Lehninger, Principles of Biochemistry (Fifth Edition). W.H. Freeman and Company, New York.

Primarose S.B. and Twyman (2003), Principles of Genome Analysis (3rd Edition). Blackwell Publishing, MA, USA

Sanger F, Nicklen S, Coulson AR (1997), DNA sequencing with chain-termination inhibitors. PNAS USA **74:** 5463-5467.

Watson, J. D. and F.H.C. Crick (1953), Molecular Structure of Nucleic Acids. Nature, **171**: 737-738.

3

Genome Assembly and Finishing

In large genome sequencing programmes the sequencing is performed from both 3′ and 5′ directions and millions of pieces of approximately 500 base pairs in length are generated from each organism. Assembly of the sequences using computational tools is a very difficult task due to the complex nature of the genomes of higher organisms. Therefore, specific software are used to perform genome assembly.

GENOME ASSEMBLY

Genome assembly is the comparison of each sequence read to every other and put them in proper order based on their overlaps. It results in collection of correctly ordered big genome stretches.

The analysis of DNA sequences starts once these are out of the sequencing machines (Fig.1).

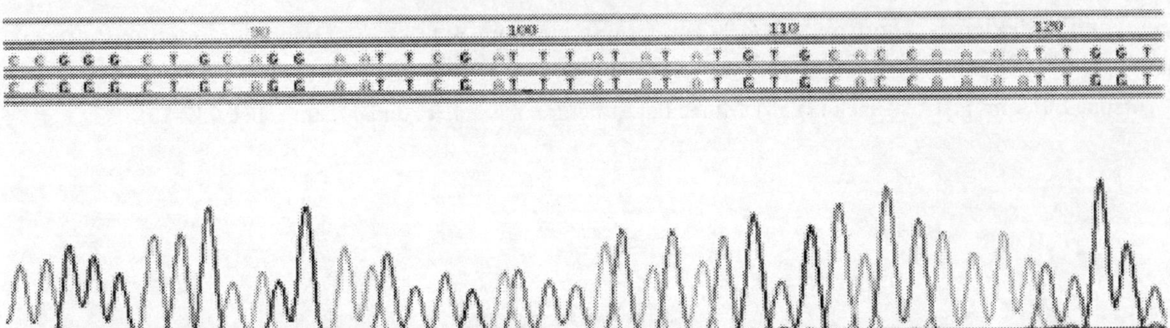

Fig. 1. DNA sequence trace file obtained from the sequencing machine.

The first and foremost task of a biologist is to look for the accuracy of sequence he obtained from the machine. It can be determined by:

- Finding cloning sites of inserts in the sequencing vector
- In case of PCR product one should look for the primer sequences used for the amplification of that product
- Performing BLAST search against related genome sequence databases and looking for probable matches.

If the unknown sequences show hits with any sequence of the same or related organisms, then it is considered a true sequence. These are the basic steps, which can be performed manually if the dataset is very small or if one has to deal with single or a few sequences. However, in large genome sequencing projects, one has to handle thousands of sequences at a given time. Hence, various bioinformatics tools are required for performing all these steps.

SOFTWARE USED FOR GENOME ASSEMBLY AND THEIR APPLICATIONS

Many software are available at different genome centers for the assembly of large genome sequences. The important software are listed in Table 1.

Table1. Name of the important assembly software and their developers.

Name of the Software	Developer
Phrap	University of Washington, USA
TIGR Assembler	The Institute for Genomic Research, USA
Celera Assembler	Celera Genomics, USA
Phusion	Sanger Cenre, UK
Atlas	Baylor College of Medicine, USA

Most commonly used software package which is freely available to the public institutions is Phred, Phrap and Consed. The brief description and use of these software is given below.

Trimming vector sequences

Once sequence data comes out of the machine it cannot be directly used for assembly purposes due to the presence of vector bone sequences in some of the trace files. Therefore, first of all, a software called cross match is used for the trimming of vector sequences. The cross match software used to compare sequence reads against the database of vector sequences and an output of vector - masked reads is obtained (Fig. 2A & B). These sequences are now ready for further analysis.

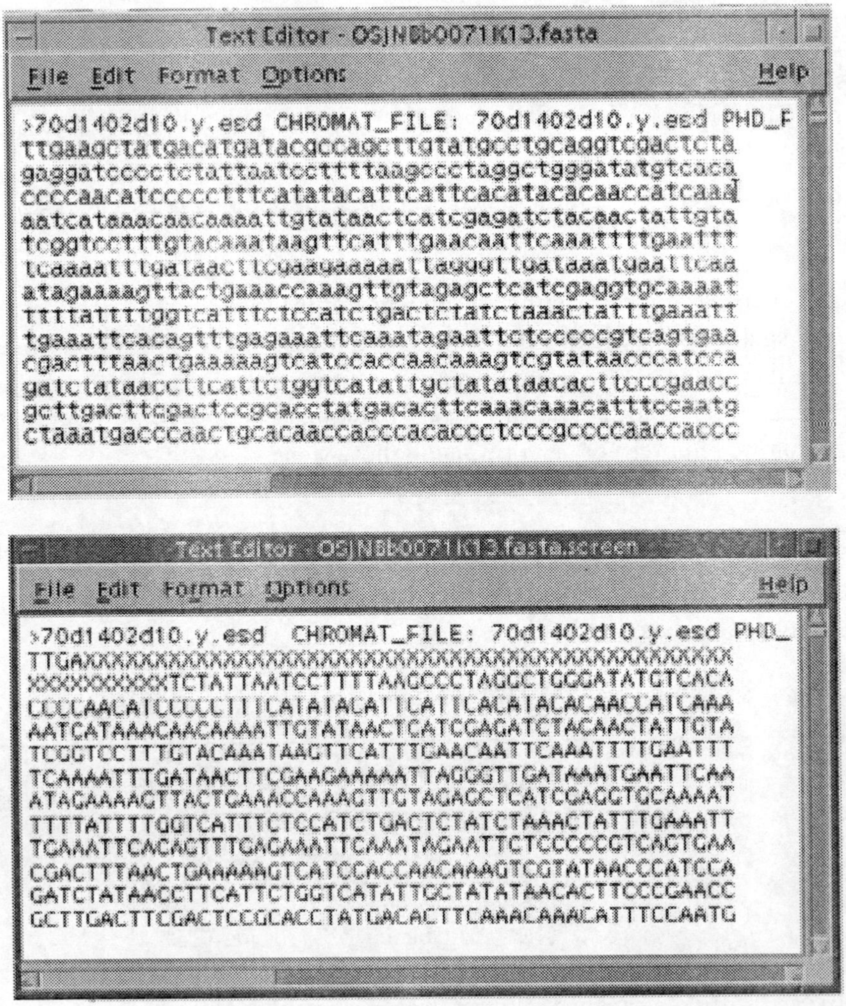

A

B

Fig. 2. DNA sequence before (A) and after (B) running cross match. XXXXX in case of B shows the vector sequences trimmed by the software.

Determination of sequence quality

The quality of sequence is of great significance when decoding complete genome of an organism. Most of the efforts in functional and comparative genomics are based on the genome sequence data available in the public databases like NCBI. Therefore, accurate analysis of raw sequence data is very important. For sequence quality analysis, the commonly used software is Phred. Each nucleotide in the trace file of the sequence is given a particular statistical score based on the trace quality in the chromatogram by Phred software. This assignment of a particular numerical value to the base is called base calling. Phred is a base calling software which assigns a particular probability score to individual base in the trace file (Fig. 3). This programme has been used very extensively in most of the large genome sequencing projects due to the high base calling accuracy which ultimately gives high quality consensus sequence. The programme has been developed by Phil Green and Brent Ewing of USA and PHRED means PHil's Read Editor. Phred takes DNA sequence chromatogram file and analyzes the peaks to call bases and then assign a quality score to each base. Four basic steps are used by this software for data analysis.

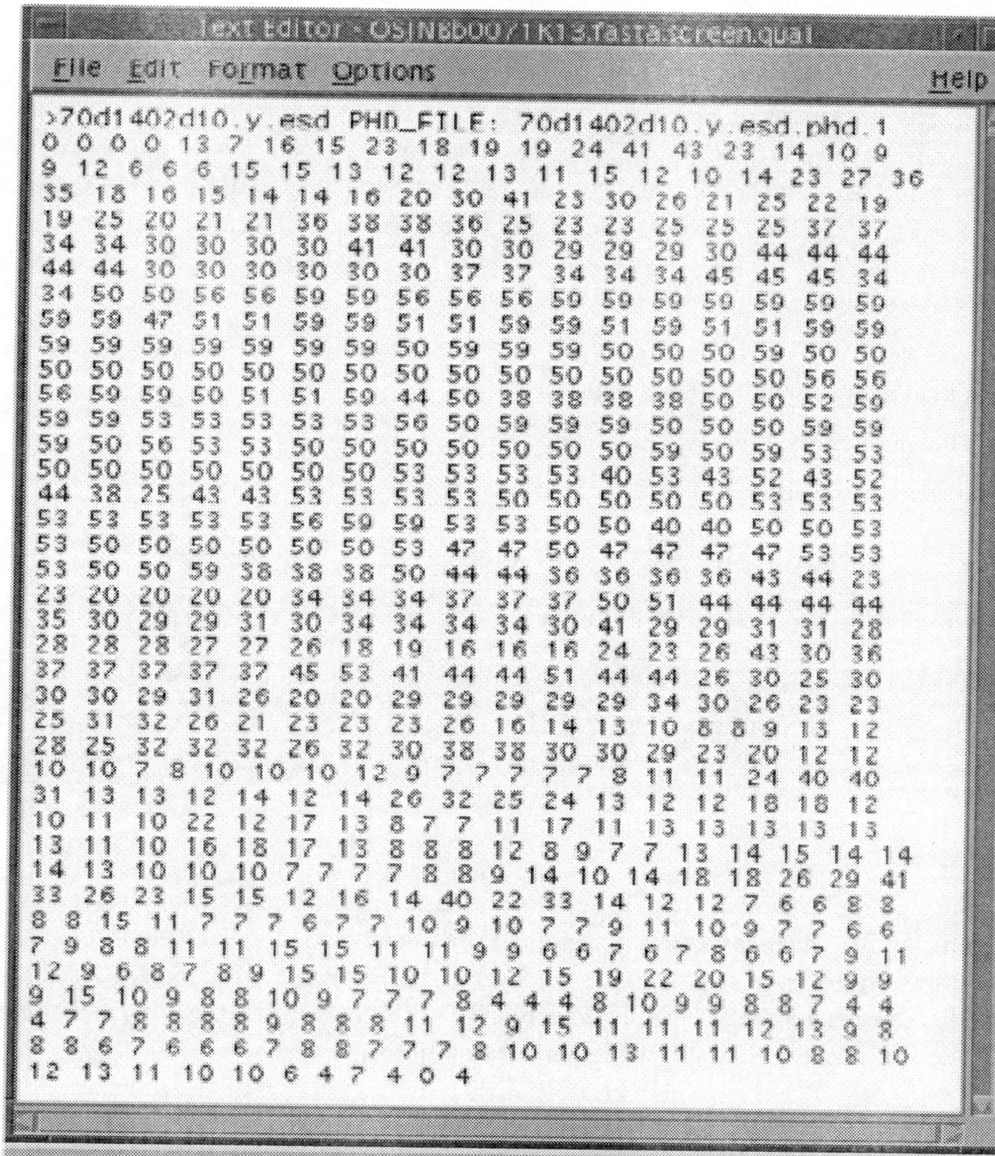

Fig. 3. Output file from the Phred showing quality score of each base.

1. It uses best region of the peak to predict expected peaks.
2. Observed peaks are identified in the chromatogram.
3. It compares both observed and expected peaks and separates all matched and unmatched peaks in two groups.
4. Software again looks at the unmatched peaks which could be called and otherwise missed in above step.

The peaks in the chromatogram files are analyzed. The peak height, peak spacing and peak compression are the basic criteria for assigning a quality score. The input file for Phred are SCF (Standard chromatogram files) or ABI trace file whereas the output files are phd files which contains Phred score of the base called (Fig.3). The quality score ranges from 4 to 60. Higher Phred score means high quality of the base. Phred score is called based on the following formulae i.e

$$q = -10 \times \log_{10}(p)$$

Where

q is the quality score i.e. Phred value

p is the probability of a base being called wrong

The meaning of different Phred scores is given in Table 2.

Table 2. Description of the Phred scores and per cent accuracy.

Phred score	Probability of wrong base call	Pre cent accuracy
40	1 in 10,000	99.99
30	1 in 1,000	99.9
20	1 in 100	99
10	1 in 10	90

Sequence assembly

One of the most popular DNA assembly software in large genome sequencing projects is called Phrap which is Phil Greens assembly program. It has been very popular among the scientific community due to the (i) fast assembly of the genomes, and (ii) it gives accurate consensus sequences. It has an ability to assemble large genome projects and can also help in identification of the sequence repeats. Basic steps used by Phrap for sequence assembly are:

1. This programme uses Phreds quality scores to identify highly accurate consensus sequences
2. All individual sequences are examined at a given position
3. The consensus sequence is built based on the highest quality sequence.

The input files for Phrap is the phd files created by Phred software and output files are known as ace files i.e. assembly files. The progamme also assigns a specific score to the bases in consensus sequence which is known as Phrap score. Phrap score is the sum of Phred quality of the overlapping sequences. For instance, if two bases are aligned and q = 15 in one of the base and q = 25 in the other base, then their sum i.e. 40 would be the Phrap score. The Phrap score 40 means 99.99% accuracy of the bases in consensus sequence. It is considered the best in high throughput genome sequences.

Assembly view

Assembly viewing and editing is performed by another graphical software known as Consed (Fig. 4). This software has been developed by David Gordon in Phil Green's laboratory and is distributed free to the academic users. Consed require three different input directories (dir) for its functioning.

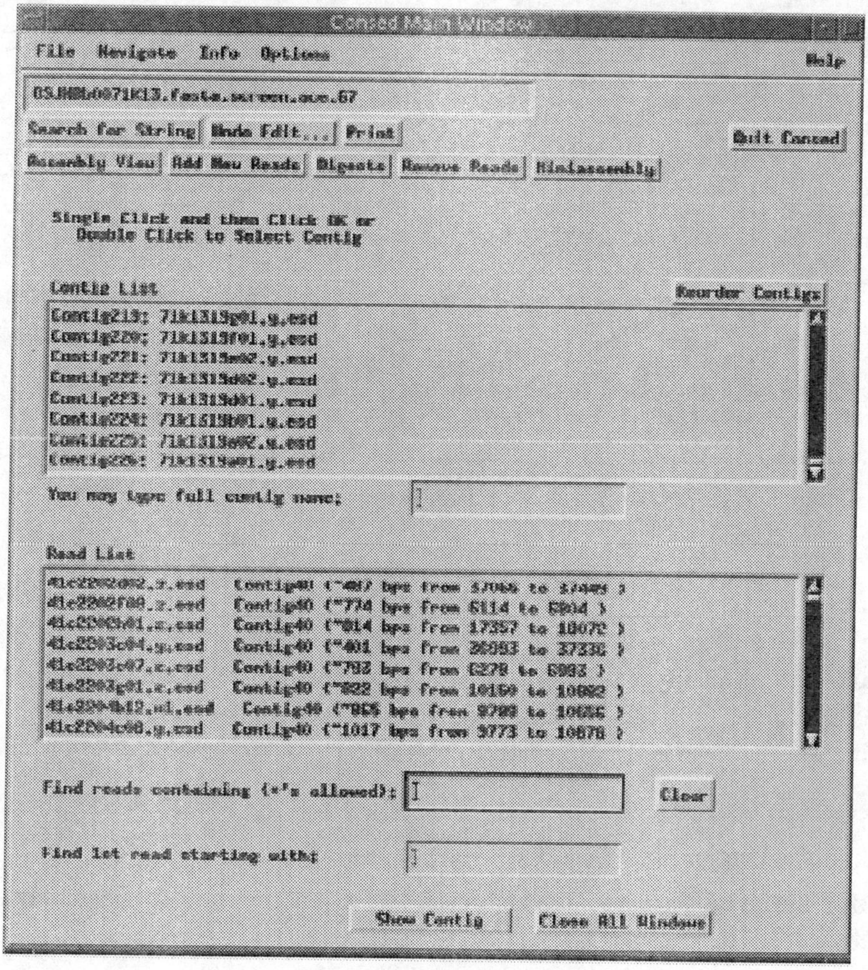

Fig. 4. Consed Window.

- chromat_dir i.e the directory of chromatogramme files generated by the sequencing machines (raw sequences).
- phd_dir i.e the output file generated by Phred software containing the quality score of each base.
- edit_dir .i.e. the directories which contain ace files generated by the Phrap software. The sequence assemblies can be viewed as a graphical interface (Fig. 5).

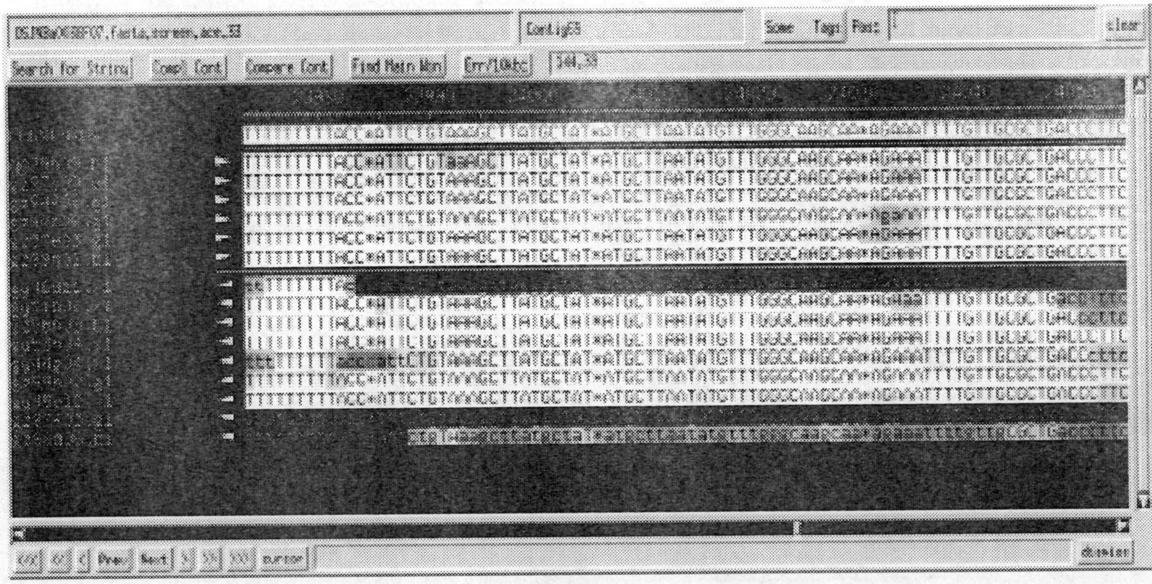

Fig. 5. A view of Consed window showing assembled sequence file.

The output of the Consed shows certain sequence quality tags. The white upper case means high quality and mismatch tags are indicated in red. Some of the important functions of Consed are that one can navigate the high quality and low quality discrepancy, compare and merge contigs of sequence data and also pick primers for finishing. Once, first assembly of the BAC clone is over, the sequences can be submitted in the GenBank in three different phases of sequences as described below and also shown in Fig. 6.

- Phase I: all the contigs above 2 Kb in size and un-oriented.
- Phase II: if the contigs are oriented and inser-vector junctions are delineated. It is also called the draft sequence.
- Phase III: the high quality sequence without any error or breaks.

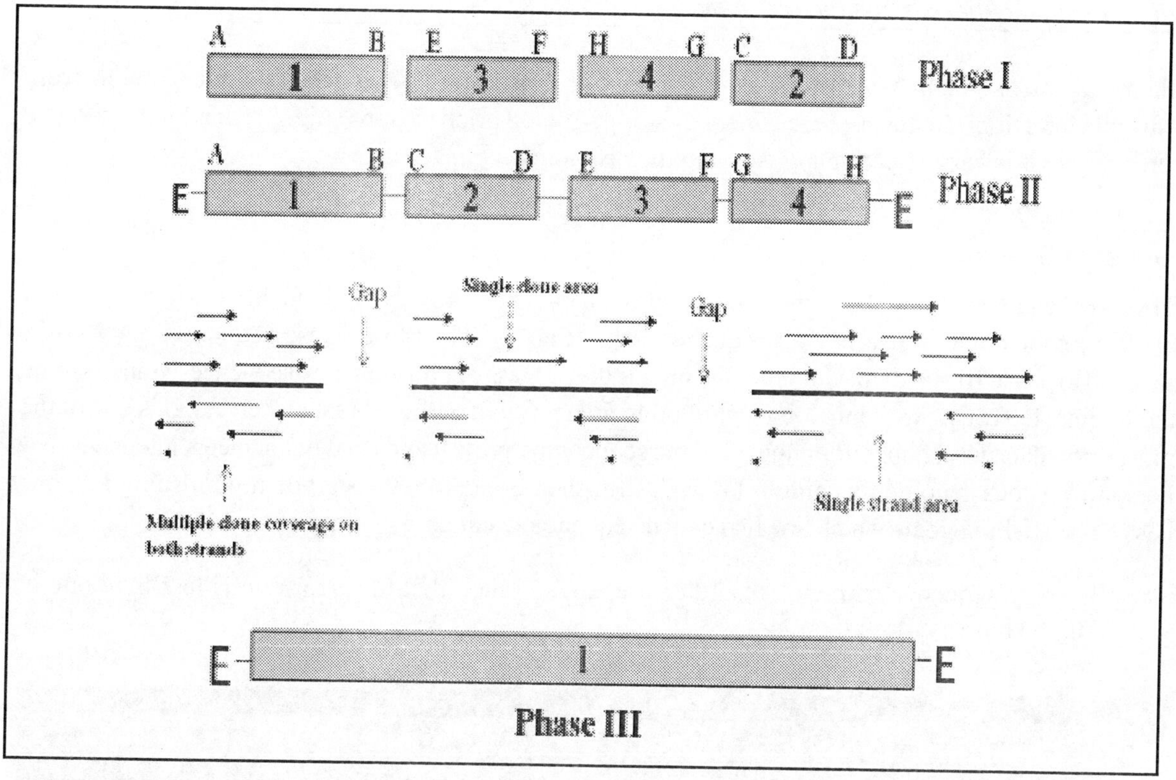

Fig. 6. Different phases of sequences submitted in the GenBank. Problems in the draft sequence (Phase11) are shown as arrows. Central solid line between arrows is the consensus sequence.

Common Problems in the Draft Sequence and Genome Finishing

Physical gaps

The physical gaps may results from the no representation of any of the sub-clones due to cloning bias while making shotgun libraries or the presence of repeat in the target regions. The physical gaps between two contigs are generally filled with PCR. By using CONSED, primers are designed from the ends of adjacent contigs and used for amplification of desired PCR product by using BAC DNA as a template. Different enzymes can be used for PCR amplification. However, Platinum *Taq* with PCR enhancer (Invitrogen) and Klentaq (Sigma) generally gives better results.

Sequence gaps

Sequence gaps are the results of poor quality sequence available at a particular point in a contig. Primer walking generally closes the sequence gaps. This is due to the compression in the sequence, mono/polynucleotide runs, which prevent extension of sequence reads. These problems can be overcome by using different sequencing chemistries like dye primer chemistry, dGTP (Perkin Elmer) chemistry or use of *in vitro* transposon containing a priming site which allows extension from multiple insertions. Sequences are generally stopped at G/C rich regions (Fig. 7) for which the dGTP terminator kit is very effective in resolving the problems.

Genome finishing

The draft sequence (Phase II) of the genomes though having relatively good quality, suffers from poor quality regions and is also having sequence gaps (Fig. 6). Therefore, these sequences have to be finished to phase III standard. Genome finishing is the process of polishing raw sequences, transforming the fragmented rough draft into long, continuous final product without breaks or errors. It is one of the most difficult tasks in high throughput genome sequencing project and is labour and cost intensive. The person who does genome finishing is known as finisher. So, before the start of any finishing job, one has to fix finishing goals which are also called sequencing standards. These are:

- Resolve sequence ambiguities and discrepancies, such that the error rate is less than one in 10,000 bases.

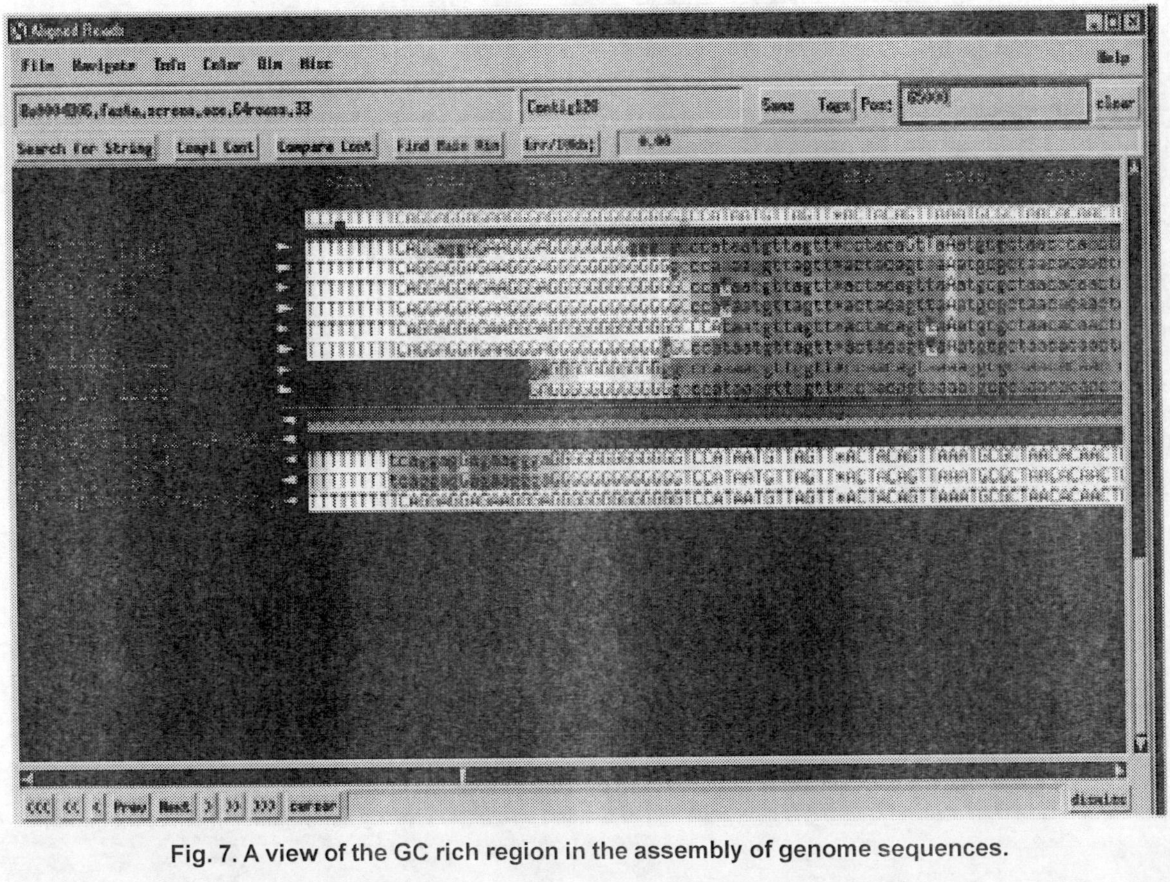

Fig. 7. A view of the GC rich region in the assembly of genome sequences.

- Provide "double-stranded" coverage for every base:
  - minimum of two different clones
  - two different directions
  - two different chemistries
- Achieve contiguity.
- Delineate vector/insert junctions.

Different methods used for genome finishing

The automated sequence editor, Consed, is used to perform sequence editing and finishing. Each contig is manually edited by navigating high quality (Phred score = 30) discrepancies > 5 bp and for any mismatch. The mismatches are tagged and evaluated from their trace files. Depending upon the type of sequence discrepancies, finisher has to design appropriate strategies.

Finisher has to scan assembly to perform following functions:

- Pick linker clones for Tn sequencing.
- Design custom oligo dye terminator.
- Design oligos for reverse dye terminator.
- Work for special chem (dGTP) reactions
- Design custom oligo for BAC DNA sequencing.
- Pick primers for PCR amplification of problem areas.

1. Transposon method

First of all whole project (draft sequence of a specific BAC clone) is scanned manually by using Consed software to find, (i) low sequence coverage areas and (2) bridge/linker clones (read pairs or mate pairs) spanning between two contigs (Fig. 8). Once the shotgun clones are identified, their DNA is used as a template for transposition. Equal amount of 8 templates (bridge clones) is pooled in such a way to get final concentration of DNA as 80 ng per pool. However, two clones can also be used in each pool for transposition. For transposition, GPS –1ve Genome Priming System (New England Biolabs Inc) can be used as per protocol supplied by the manufacturer. Basic steps used in transposon methods are listed below:

- Identify linker clones (automated/manual methods).
- Perform transposon insertions.
- Transform DH10B cells.

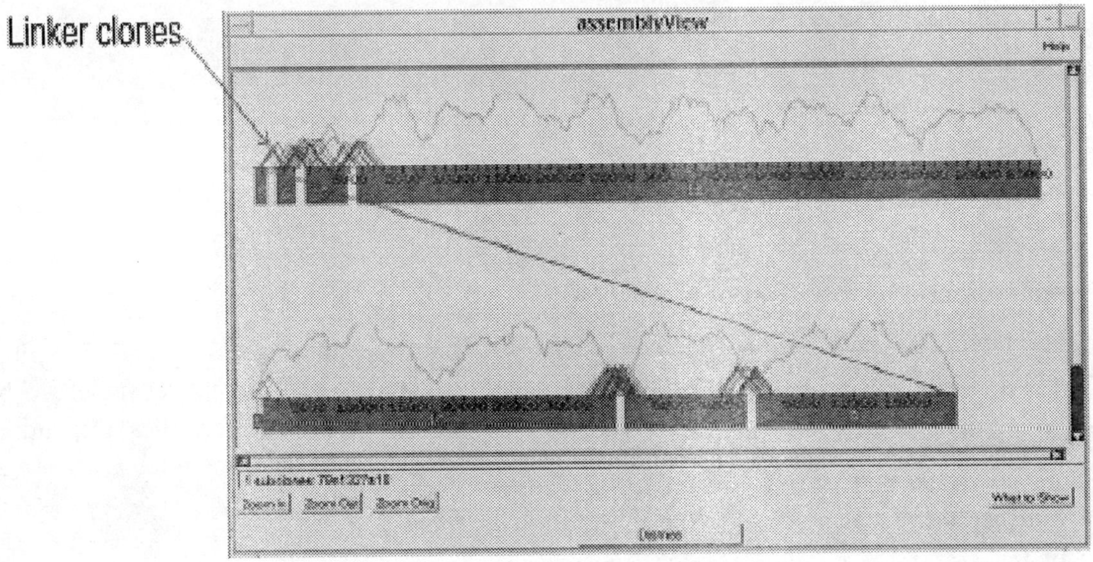

Fig. 8. Identification of Linker clones (pair reads) which joins two contigs by using consed software.

- Pickup atleast 24 white colonies.
- Prepare template.
- Sequencing of all the templates.
- Add new sequence data in the assembly.

It will help in joining most of the contigs. After this, second round of finishing starts again by navigating individual sequence contigs using Consed softwares.

2. Custom primer method

After first round of editing, second navigation round is performed by looking for high/low sequence discrepancies, assembly with gaps or having poor sequence data at the joints, sequences with less than two high quality reads at a particular location and poor quality bases etc. Poor quality bases are manually edited wherever possible by looking into the trace files otherwise the clone is selected for re-sequencing. For filling sequence gaps, custom primers are designed with the help of automated primer picking preferences of the Consed. The basic criteria for picking primer should be that the length of primer be about 20 nt, GC contents, 40-60% and Tm 60 % etc so that more specific sequencing can be performed. Then depending upon the type of poor region, the custom primers are designed for primer walking by using template of the clones spanning to the region (Fig. 9). In this case, instead of universal sequencing primers, custom primers are used for sequencing. The composition of sequencing reaction used is Big Dye(v3.2, Perkins Elmer) 1.5 μl, dGTP 0.5 μl and sequencing buffer (Perkin Elmer) 2 μl. Template (5 μl) and custom primer (10nm) 1.0 μl in a total reaction volume of 10 μl are added to the sequencing reaction. In the GC rich region, dGTP chemistry (Perkin Elmer) can be used. The data obtained after sequencing is added to the project directory by using Add New Reads option of the Consed.Various steps used in this method are given in Fig. 9.

3. PCR method

The PCR method is used to resolve physical sequence gaps within the contigs. These primers can also be designed with the help of PCR primer design option available within the Consed software. Here again the basic criteria used for primer has to be followed besides keeping primer length more than 24 nt so that very specific amplifications can be achieved. Once primer design and synthesis is over, these are used for the amplification of BAC DNA with high fidelity DNA polymerase. Then, PCR products are sequenced after cleanup. The data obtained after sequencing is added to the project directory by using Add New Reads option of the Consed. All the steps followed in this method are given in Fig. 10.

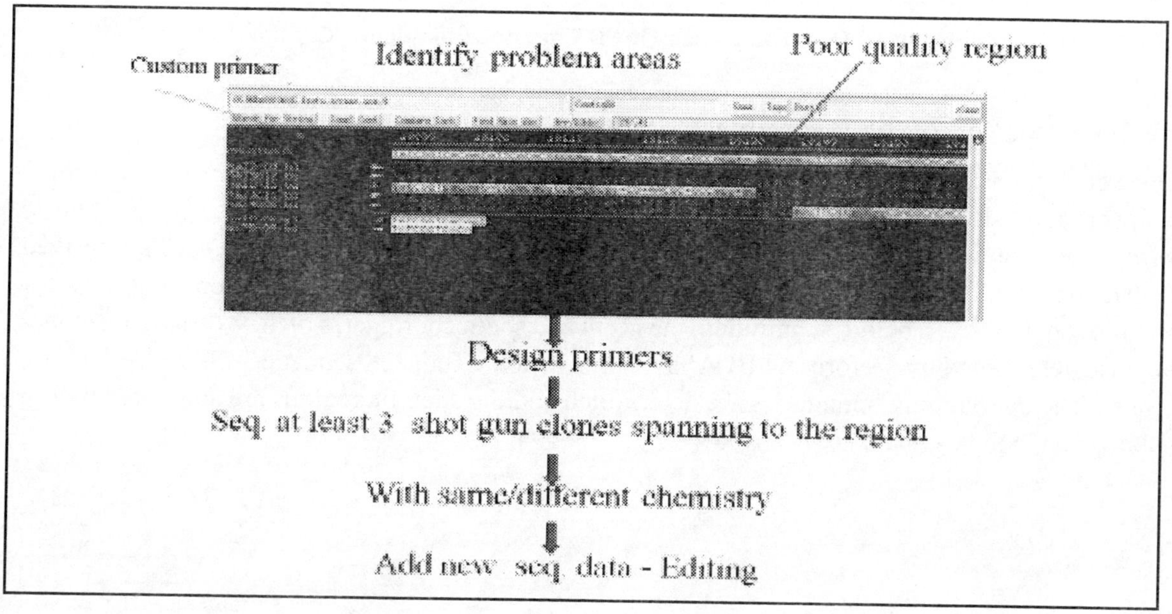

Fig. 9. Designing custom primers for finishing a Contig and the steps used in editing poor regions.

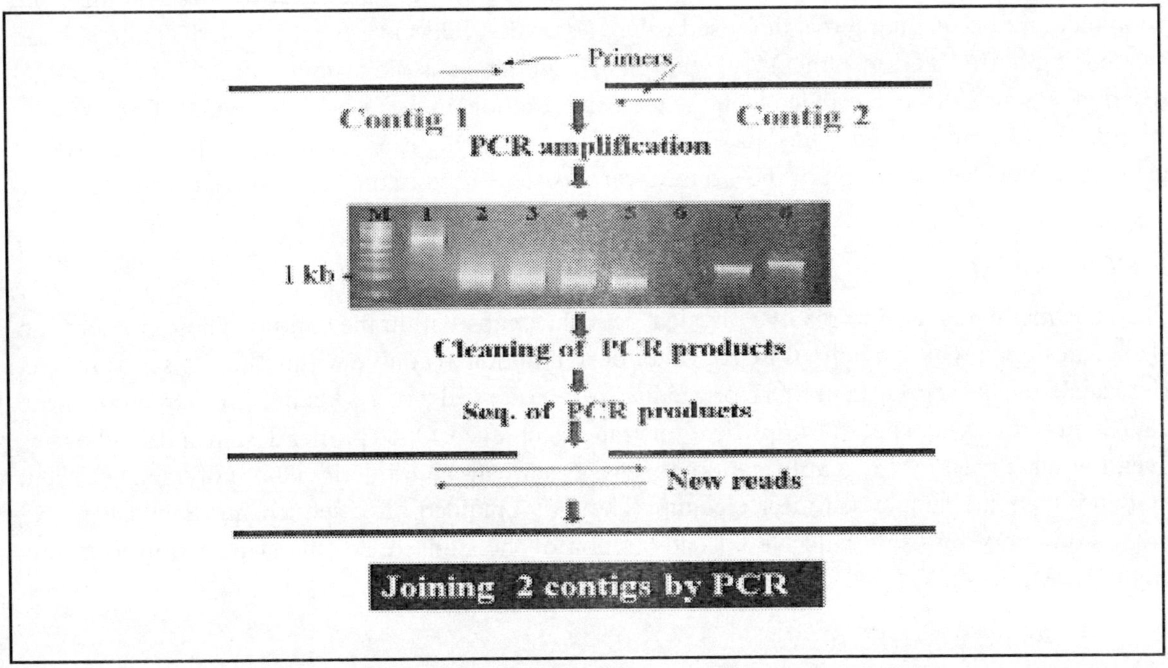

Fig. 10. Different steps used in PCR methods for joining Contigs.

Final Verification of the Assembly

Once the clone is finished (Fig. 11), whole sequence is then verified for its accuracy, by performing Mapsort analysis i.e. sequence is digested *in silico* with at least 3 restriction enzymes. The *in silico* digests are compared with the actual fingerprints of the BAC clone. If *in silico* derived restriction digests and fingerprints of the BAC clone matches well, then clone is ready for submission. However, before submission, the north and south coordinates of the overlapping clones are also determined by performing BLAST search with the finished sequence of the overlapping clones. The overlapping boundaries are determined and the specific regions are annotated before submission to the Genbank.

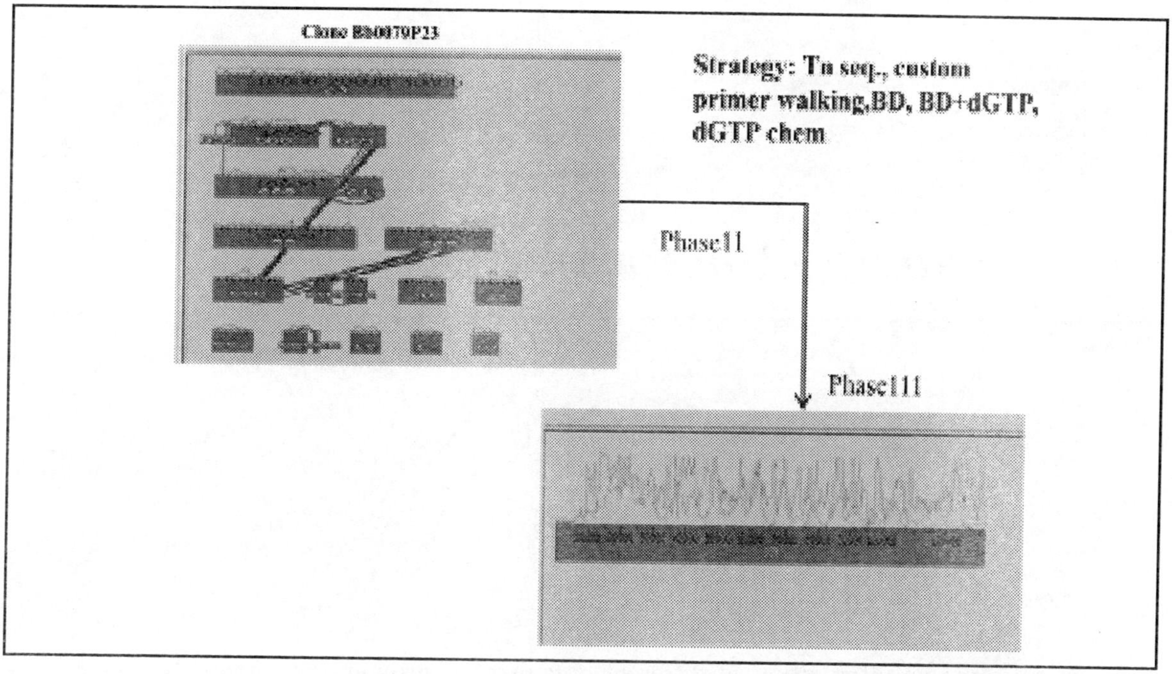

Fig. 11. Problems in the clones which is in different contigs in phase II and same clone in phase III after finishing.

Suggested Reading

Brent E. and P. Green (1998), Base-calling of automated sequencer traces using phred. II. Error probabilities. Genome Res. **8**: 186-194.

Brent E , H. LaDeana, C. M. Wendl, and P. Green (1998), Base-calling of automated sequencer traces using phred. I. Accuracy assessment. Genome Res. **8**: 175-185.

Chaisson M, P. Pevzner and H. Tang (2004), Fragment assembly with short reads. Bioinformatics **20**: 2067-2074.

Gordon D, C. Abajian and P. Green (1998), Consed: a graphical tool for sequence finishing. Genome Res. **8**: 195-202.

Mullikin J.C. and Z. Ning (2003), The phusion assembler. Genome Res. **13**: 81-90.

Pop M and D Kosack (2004) Using the TIGR assembler in shotgun sequencing projects. *Methods Mol Biol.* **255**: 279-94.

4

Genome Databases

The genome sequence information of different organisms is growing exponentially. Systematic storage and curation of this information is very important so that it can be used for the welfare of mankind. The sequence information is a life long treasure which should be kept in safe places in the public databases.

WHAT IS A DATABASE?

A database is a collection of information stored in a computer in a systematic way, such that a computer program can consult it to answer questions. A biological database is a large, organized body of persistent data, usually associated with computerized software designed to update, query, and retrieve components of the data stored within the system. A simple database might be a single file containing many records, each of which includes the same set of information. For example, a record associated with a nucleotide sequence database typically contains information such as contact name; the input sequence with a description of the type of molecule; the scientific name of the source organism from which it was isolated; and, often, literature citations associated with the sequence.

TYPES OF DATABASES

1. Flat-file databases
2. Relational databases
3. Hierarchical databases

1. Flat-file database

The flat-file databases are the earliest and simplest databases which are mainly used for storing small amount of any types of data. These databases are made up of a set of strings in one or more files that can be parsed to get the stored information. A common delimit is used to split up a flat-file database. For very simple data it may be a comma however for more complex strings, tabs, new lines or combination of characters are used.

Advantages

It is easy to set up and understand flat file database.

Disadvantages

- It may require entering the same information in many records.
- It is hard to read a text database.
- It may render the file unreadable and un-editable to anyone looking after the database due to the complex storage methods.

2. Relational database

Relational databases use a set of tables to organize the data. The database is made up of different tables and each table consists of rows and columns. The columns represent individual fields. These columns are indexed according to common features known as attributes for linking to other tables. [The rows in a table values in the field of records called as tuples.] For extracting information from a relational database, the system collects the linked items from different tables, combines them and presents in the form of a single report. The relational database is constructed by using a programming language called structured query language (SQL). Before developing relational database it is essential to develop well defined architecture (Fig. 1) and design of the database.

Advantages

- Duplication of data entry can be reduced.
- Very fast search for getting information in a single report.
- Can create queries to answer complex questions.

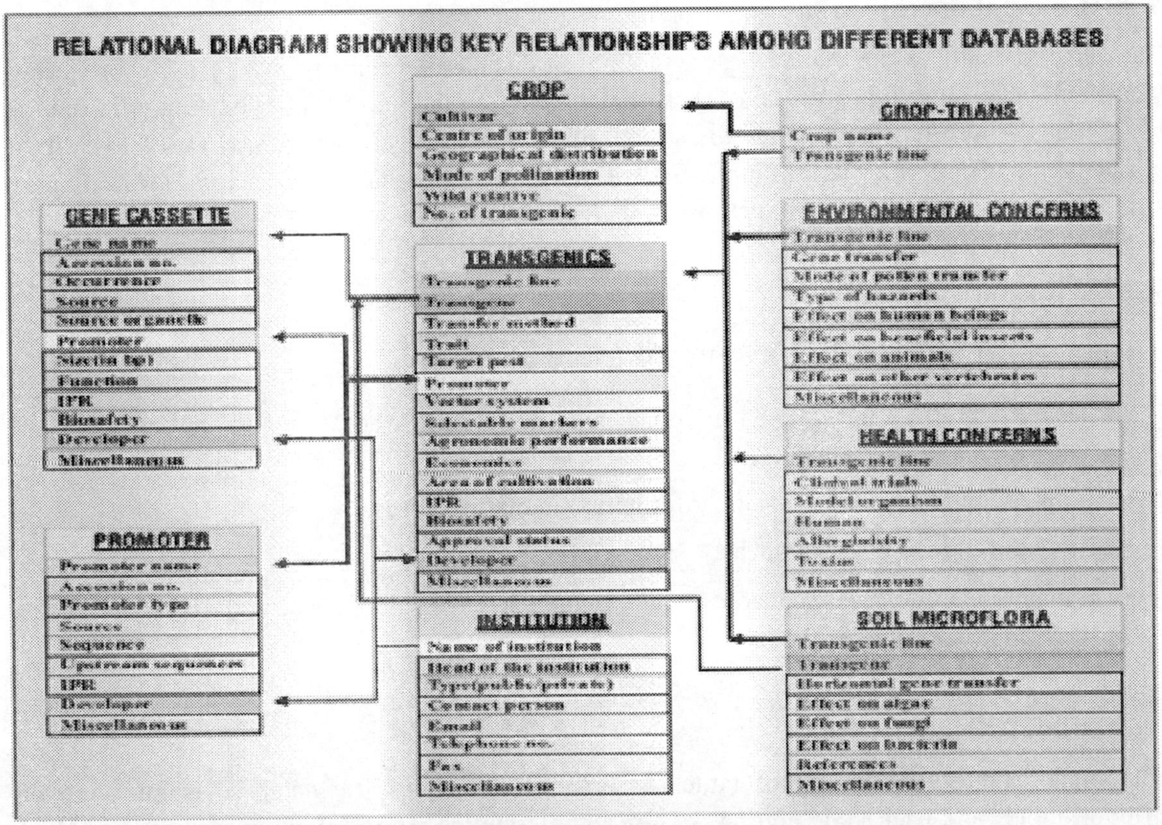

Fig. 1. An example of the architecture of GM crops database.

Disadvantages

- It uses many tables hence complex to set up.
- Relationship between each part is difficult to understand.
- Need more intellectual input while designing the architecture.

3. Hierarchical database

In this type of database the data is organized in a typical hierarchical or ordered tree structure. The nodes of the tree represent record type. Each tree has a well defined root record type which can be referred as level 0. Therefore, the record type which depends on level 0 are known as level 1 and the dependent on level 1 is known as level 2 and so on as shown in the Fig. 2.

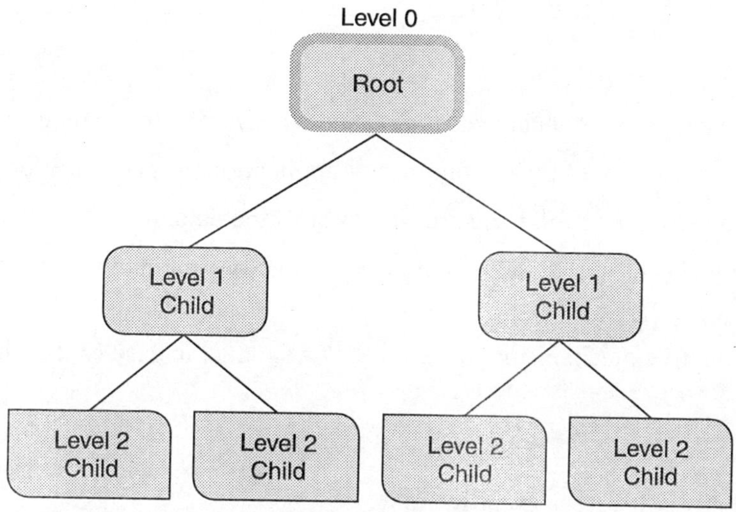

Fig. 2. Hierarchical database model.

Advantages

- The construction and operation of hierarchical model is simple.
- It is very easy to add or delete the records from the database.
- Data retrieval is fast.
- Database construction language is also simple.

Disadvantages

- It is a navigational type of database hence more time consuming.
- Repetitive data and separate record is required for each association.
- It needs more storage space.

Database Management System

Database management systems (DBMS) are the computer software programs which facilitate the access and retrieval of information by systematically organizing, searching, and accessing data. Beside containing raw data records, the DBMS also have the operational instructions to help identify hidden connections among data records. These DBMS can be classified into two forms depending upon the types of data structures. These are relational database management systems and object-oriented database management systems.

Advantages

- It helps in reducing the redundancies of data and amount of storage required.
- The database can be shared by any number of application program or users.
- Due to the centralized control, it ensures integrity of the data.
- Proper security checks are implemented in the DBMS for maintaining confidentiality of the data.
- DBMS has effective mechanisms for recovery of the data in case of its failure.

Disadvantages

- It involves costly software and hardware.
- Data backup and recovery is very complex.
- Centralized database has it own associated problems.

RELATIONAL DATABASE MANAGEMENT SYSTEM

Relational Database Management System (RDBMS) is used as one of the most commonly used DBMS. An RDBMS is a program that lets you create, update, insert, retrieve and administer a relational database. The RDBMS used for explanation of this chapter is MySQL. Since, MySQL is a full-fledged open source relational database management system, various steps used for the creation of a database using MySQL are explained in following sections.

1. Data Processing

Data processing is any process that converts data into information. All the information are collected from relevant papers available on net. Then these collected raw data is parsed into useful information, so that it can be stored in MySQL database. Hence, after processing the database is stored in MySQL database (backend).

2. Architecture used for application development

A three-tier architecture can be used in a typical database, which includes **Client Tier** or user interface, **Middle Tier** or business logic and **Data Storage Tier** (Fig. 3). Applied to web applications and distributed programming, the three logical tiers usually correspond to the physical separation between three types of devices or hosts. These devices are **Browser** or **GUI (General Users Interface) Application, Web Server** or **Application Server** and **Database Server** (RDBMS).

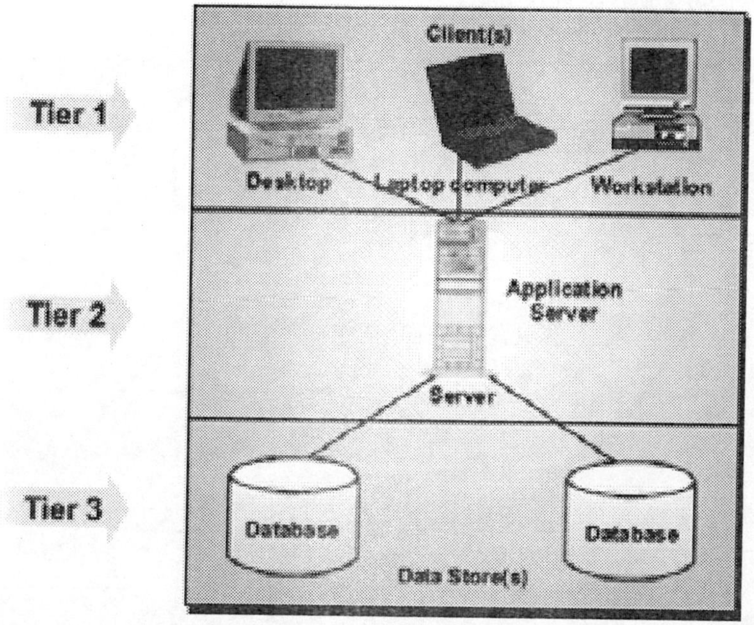

Fig. 3. The schema of a three-tier architecture used for the development of RDBMS. In this configuration, the application server provides authentication services, database connection services, and application processing services. The client's role is to initiate the request and display the results returned, while the database serves as the repository for the data.

3. MVC Architecture

Model View Controller (MVC) is architecture for building applications that separate the data (model) from the user interface (view) and the processing (controller). MVC is widely used in Web-based applications (Fig. 4). MVC is a software architecture that separates an applications data model, user interface and control logic into three distinct components so that modifications to one component can be applied with minimal impact to the others.

WORKING OF MVC ARCHITECTURE

The original request in MVC is handled by a servlet. The servlet invokes the data-access code and creates beans to represent the results. Then, the Servlet (the controller) decides which Java Server Page (JSP view) is appropriate to present those particular results and forwards the request there. The servlet decides what business logic code applies and which JSP page should present the results. Following steps are involved in MVC architecture.

- It defines beans to represent the data.
- It uses a servlet to handle requests.
- Populate the beans.

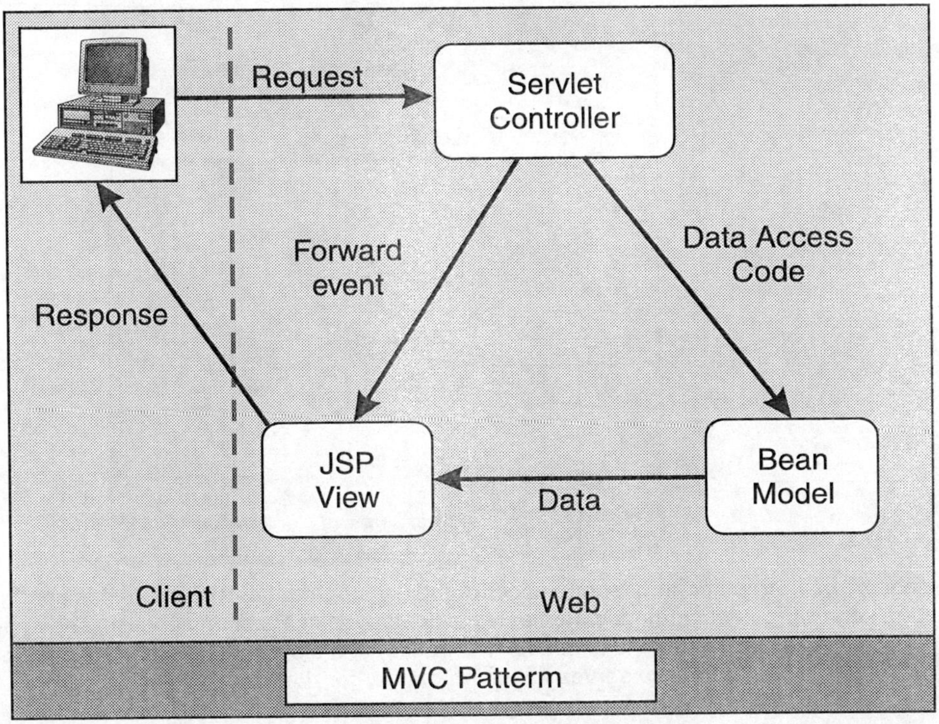

Fig. 4. The schema of MVC architecture.

- The servlet invokes data-access code to obtain the results. The results are placed in the beans.

- Store the bean in the request session or servlet context.
- Forward the request to a JSP page.
  - The servlet determines which JSP page is appropriate to the situation and uses the forward method of **Request Dispatcher ()** to transfer control to that page.
- Extract the data from the beans.
  - The JSP page accesses beans with **jsp: useBean** and a matching scope.

The brief description of different components is given below.

- **Servlet**: A Java application that runs in a web server or application server and provides server-side processing such as accessing a database and e-commerce transactions. Since they are written in Java, servlets are portable between servers and operating systems. Servlets receive and respond to requests from web clients, usually across HTTP, (Hyper Text Transfer Protocol).

- **Java Server Page**: It is an extension to the Java servlet technology from Sun microsystems that allows HTML (Hyper Text Markup Language) to be combined with Java on the same page. The Java provides the processing and the HTML provides the page layout that will be rendered in the Web browser. JSPs are compiled into Servlet by a JSP compiler. A JSP compiler may generate a servlet in Java code and then compiled by the Java compiler, or it may generate byte code for the servlet directly.

- **Java beans**: These are the reusable software components. Beans are simply Java classes that are written in a standard format. Java Beans are independent Java program modules that are called for and executed. These are used primarily for storing data.

- **Java Data Base Connectivity (JDBC)**: It is a programming interface that lets Java applications access a database via the SQL language. Since Java interpreters (Java Virtual Machines) are available for all major client platforms, this allows a platform-independent database application to be written. The JDBC provides a standard library for accessing relational databases. It provides methods for querying and updating data in a database.

The basic flow chart of the GM crops database developed at NRCPB, New Delhi is given in Fig. 5. This figure showed the flow of information from a desktop from where a query is fired on the servers, which in return provides required information in user's friendly manner.

BIOLOGICAL DATABASES

With the advent of genomic research enormous amount of raw protein or DNA sequence data are being generated from different biological organisms. To handle this huge biological data, sophisticated computational methodologies are required to manage and utilize it effectively. Therefore, for proper storage, curation and retrieval of biological data, it is of utmost importance. The development of databases is required to store molecular biological information in a systematic manner. Different types of biological database are given in Fig. 6. These databases are specifically developed for protein or DNA sequence storage or for both under different divisions.

i) Divisions of DNA databases

Since the size of databases is growing rapidly, these have been further broken into divisions on the basis of the taxonomy of the organisms. The GenBank divisions are divided into two general categories like, organismal and functional categories. The sequences derived from specific organisms are stored in the organismal category. Whereas the functional category includes databases which are independent of their taxonomic classification e.g. EST, STS and HTG etc. Respective Genbank divisions store sequence records of different organisms which are identified from three letter codes indicated in the beginning of each sequence entry. For instance, HTG

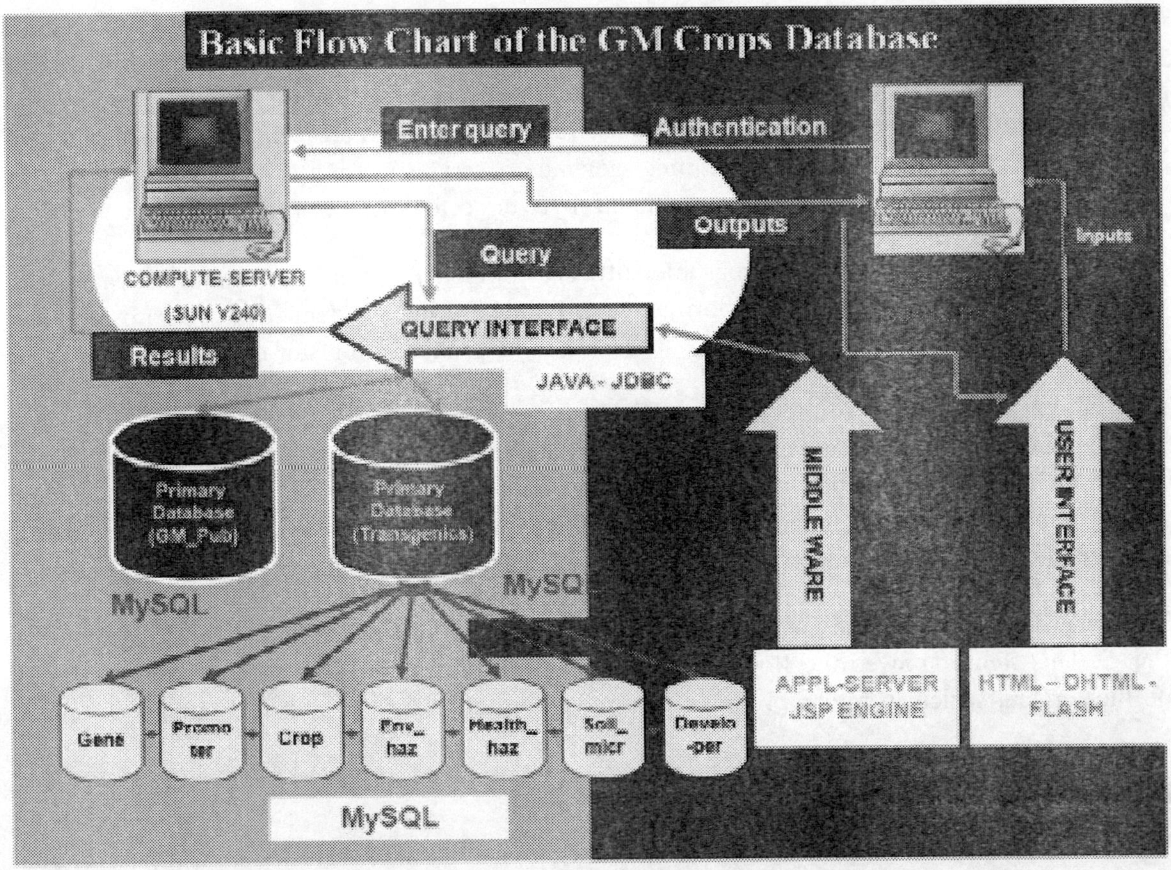

Fig. 5. Basic work flow chart of a typical database constructed with RDBMS.

(high throughput genome) division contained sequences generated from different organisms using high through-put genome sequencing approach. These sequences are generally unfinished and are further classified as Phase 1(sequences which are unfinished, unordered and contain gaps) and Phase 2 (sequences which are unfinished, ordered and contain a few gaps). Once sequences are finished and all gaps are resolved (Phase 3) it is moved to a specific division e.g. PLN in case of plants.

ii) Divisions of protein databases

Protein sequences are mainly stored in both the databases i.e. EMBL and GenBank. Swiss-Prot, a very well maintained and refined database was established at the Swiss Institute of Bioinformatics. Though, it is a small database, it has important annotations which are freely available to the academic users. GenBank created PIR – a protein database as a translation of the GenBank. PIR database is further subdivided into four sections like PIR1, PIR2, PIR 3 and PIR4 on the basis of degree of annotation.

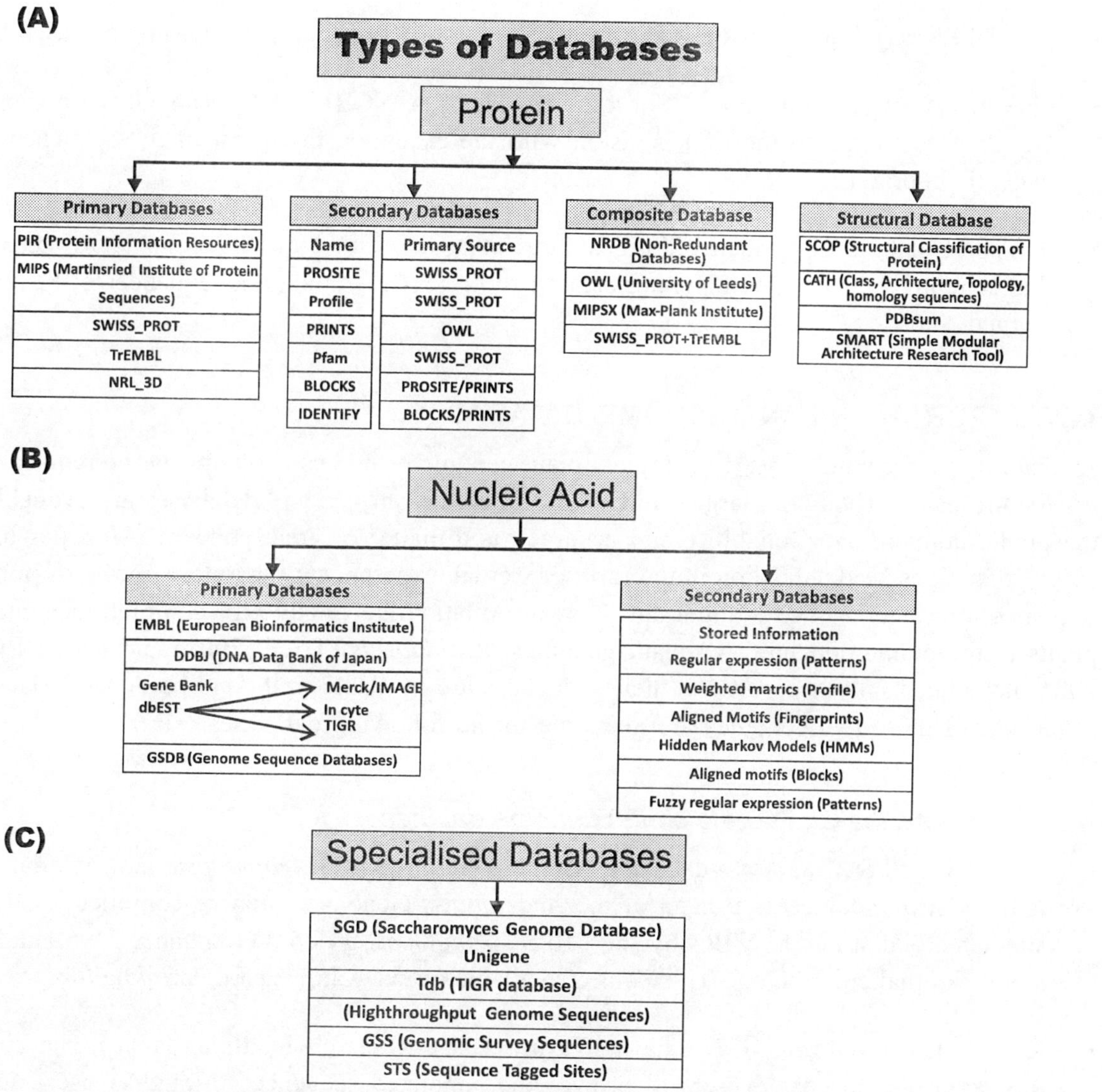

Fig. 6. Different types of biological databases. A) Protein , B) DNA and C) Specialized databases.

DIFFERENT CLASSES OF PLANT GENOME DATABASES

- dbEST (Database of Expressed Sequence Tags at NCBI, USA) is a division of **GenBank** that contains sequence data and other information on "single-pass" cDNA sequences, or Expressed Sequence Tags, from a number of organisms.

- TGI (TIGR Genome Indices, integrated analysis of public EST data, TIGR, USA) TIGR's Genome Projects are a collection of refined databases containing DNA and protein sequence, gene expression, cellular role, protein family and taxonomic data for plants.

- Mendel ESTs (Database of annotated plant ESTs in dbEST at John Innes Centre, Norwich,UK).

- dbGSS (Database of Genome Survey Sequences at NCBI, USA) The GSS division of GenBank is similar to the EST division with the exception that most of the sequences are genomic in origin.

- dbSTS (Database of Sequence Tagged Sites at NCBI, USA) dbSTS is an NCBI resource that contains sequence and mapping data on short genomic landmark sequences or Sequence Tagged Sites.

TWO IMPORTANT PLANT GENOME DATABASES

Availability of accurate or nearly complete plant genome resources is one of the prerequisites of *in-silico* research. Though, a large number of scattered plant genome databases are available in the public domain, their reliability and accuracy is a matter of great concern. All types of *in-silico* predictions need validation in the form of wet lab experiments. Therefore, purity of primary sequence information is very important. Less information is available for complete genomes of plants in the public domain. Two plant genomes, *Arabidopsis* (TAGI, 2001) and Rice (IRGSP, 2005) have been fully decoded and their sequences are available in the public domain. The full sequences of these two genomes are available in the following databases:

i) Arabidopsis thaliana (Thale cress) genome databases

- MATDB (MIPS *A. thaliana* database, Munich, Germ.) MIPS *Arabidopsis thaliana* database is the world wide access to data of the *Arabidopsis* Genome Initiative compiled, analysed, annotated and stored at MIPS by the MIPS *Arabidopsis* group and enhanced by data from many external contributors. This data is available on www.mips.gsf.de/proj/thal/db

- KAOS (Kazusa *Arabidopsis* data at Kazusa DNA Research Institute, Japan) The aim of this service is to enable users to browse the annotated sequences through a user-friendly graphic system and search engines (www.kazusa.or.jp/kaos).

- TIGR *Arabidopsis thaliana* Database (The Institute of Genomic Research, Rockeville MD, USA) TIGR has now finished the complete re-annotation of the *Arabidopsis* genome to a uniform high standard and made it available to the plant biologists around the world (www.tigr.org/tdb/e2k1/ath1/).

ii) Oryza sativa (Rice) genome databases

- The International Rice Genome Sequencing Project (IRGSP) has updated the genome sequence of *Oryza sativa* ssp. *japonica* cultivar Nipponbare with the release of the Build 4.0 pseudomolecules (www.rgp.dna.affrc.go.jp/IRGSP/). The nucleotide sequence representing the entire chromosome was constructed by joining the sequence of each PAC/BAC clone based on the order of the clones on the latest physical map. The overlapping sequences were removed and the physical gaps were replaced by successive Ns to make it in the form of 12 pseudomolecules representing all the 12 rice chromosomes.

- Gramene (comparative mapping resource for grains) is a genome information resource of important plants. The basic purpose of this database is to provide added value to the data sets available within the public domain. It will facilitate researchers' ability to understand the rice genome and use the rice genomic sequence information for identifying and understanding corresponding genes, pathways and phenotypes in other crop plants like wheat and barley (www.gramene.org).

- INE (Integrated rice genome explorer) A database integrating the genetic map, physical map and sequence information of the rice genome (http://rgp.dna.affrc.go.jp/giot/INE.html) is available in the public domain.

Beside these, the development of genome resources in other major crops like wheat, maize, barley, *Brassica*, *Medicago* etc. are given in Table 2:

Table 2: List of important plant genome databases and their web addresses

Plant Name	Database	Web Address
Arabidopsis thaliana (Thale cress)	MATDB (MIPS *A. thaliana* database, Munich, Germ.)	http://mips.gsf.de/proj/thal/db/index.html
	TAIR (The *Arabidopsis* Information Resource, previously AtDB, at Stanford, USA)	http://www.arabidopsis.org/search/
	KAOS (Kazusa *Arabidopsis* data Opening Site at Kazusa DNA Research Institute, Japan)	http://www.kazusa.or.jp/kaos/
	Arabidopsis Genome Analysis (Cold Spring Harbor laboratories, USA)	http://www.cshl.org/
	TIGR *Arabidopsis thaliana* Database (TIGR, Rockeville MD, USA)	http://arabidopsis.tigr.org/
Oryza sativa (Rice)	RGP (Rice Genome Research Programme, Japan	http://rgp.dna.affrc.go.jp/index.html
	Gramene (Comparative mapping resource for grains)	http://www.gramene.org
	INE (Integrated rice genome explorer: IRGSP, Japan)	http://rgp.dna.affrc.go.jp/giot/INE.html

Contd....

Plant Name	Database	Web Address
Triticum aestivum (Wheat)	The Grain genes database	http://www.graingenes.org/
	The ECP/GR wheat database, RICP	http://www.ecpgr.cgiar.org/databases/crops/wheat.htm
	The Field food crop (International rice corporation)	http://www.fao.org/AG/AGP/AGPC/doc/field/Wheat/data.htm
	The TIGR Wheat genome Annotation	http://www.tigr.org/tdb/e2k1/tae1/
Zea mays (Maize)	The TIGR Maize genome Database	http://maize.tigr.org/
		http://www.ecpgr.cgiar.org/databases/crops/tomato.htm
	The ECP/GR maize database, RICP	http://bioresearch.ac.uk/browse/mesh/D003313.html
	BIORES	
Lycopersicon (Tomato)	The Tomato Genetics Resource Center	http://tgrc.ucdavis.edu/
	The Tomato Expression Database	http://ted.bti.cornell.edu/
	The International Solanaceae	http://www.sgn.cornell.edu/
	Genome Project Tomato database	http://slofly.com/tomatodb/
Brassica napus	The ECP/GR *Brassica* database, RICP	http://www.ecpgr.cgiar.org/databases/crops/brassica.htm
	The European *Brassica* Databases	http://www.actahort.org/books/459/459_28.htm
	Natural Research Environment Council	http://www.brassica.info/ssr/SSRinfo.htm
Medicago truncatula	The TIGR Database	http://www.tigr.org/tdb/e2k1/mta1/
	A model for legume research	http://medicago.org/
	The ECP/GR *Medicago* database, RICP	http://www.ecpgr.cgiar.org/databases/Crops/Medicago.htm
	The *Medicago* database Query (Agricultural Research Organization of Israel)	http://bioinfo.agri.gov.il/cgi-bin/medicago_query.pl
	Medicago truncatula Sequencing Resoucrces	http://medicago.org/genome/
	Centre for *Medicago* genome Research	http://www.noble.org/medicago/index.htm
	The legume Information System (NCGR)	http://www.comparative-legumes.org/
Hordeum vulgare (Barley)	The Plants for a Future	http://www.pfaf.org/database/plants.php?Hordeum+vulgare
	MEROPS *Hordeum vulgare*	http://merops.sanger.ac.uk/cgi-bin/speccards?sp=sp000152&type=P
	TENN Vascular Plants	http://tenn.bio.utk.edu/vascular/database/vascular-database.asp

Suggested Reading

Bension, D.A., I, Karsch-Mizrachi, D.J. Lipman ***et al.*** (2005). GenBank. Nucleic Acids Res. **33**: D34-D38.

Berman H.M., J. Westbrook, Z. Feng *et al.* (2000), The Protein Data Bank. Nucleic Acids Res. 28, 235-242.

Bresson, S. and B. Catania (2006), Introduction to database systems. New York: McGraw Hill Higher Education.

Date, C.J. (1995), An Introduction to Database Systems. 6th ed. Boston: Addison-Wesley.

Wheeler D.L., T, Barrett, D.A. Benson *et al.* (2006), Database Resources of the National Centre for Biotechnology Information. Nucleic Acids Res. 34, D173-D180.

5

Pair-wise Sequence Alignment

Alignment is the process of finding maximum similarity between two or more than two protein or DNA sequences by their mutual rearrangement. If alignment is performed over the entire sequence length then it is known as Global Sequence Alignment. The alignment when performed over some particular segments of the sequence is known as Local Alignment. If alignment is performed between two sequences then it is known as Pairwise Alignment. Alignment performed among three or more than three sequences is known as Multiple Sequence Alignment. Multiple sequence alignment is generally performed for searching conserved regions among the sequences.

BASIC PROCESS OF ALIGNMENT

The process of alignment involves some edit operations in the sequence. The insertion, deletion and substitution are three main edit operations. **In**sertion & **Del**etion are commonly known as **Indels.** Substitution includes replacement of one alphabet with the same or different alphabets in either of the sequences. For aligning two sequences, we can insert spaces either within the sequence or at any other place. Then we can place two resulting sequences one above the other, so that every character or space in either of the sequences should be opposite to a unique character or space in the other sequence.

e.g. Suppose we want to align two sequences X & Y.

```
X = " A T T C G T A"
      | |
Y = " A T C T A A"
```

In case, these sequences are aligned as shown above then we can find only two matches. 'A' & 'T' of sequence *'X'* are matching with 'A' & 'T' of sequence *'Y'*, respectively. Rest 4 alphabets are mismatching for both sequences and last alphabets of sequence *'X'* is matching with a gap in the *'Y'*. We can observe that there are two matches, 4 mismatches and 1 gap in these sequences before alignment.

Let us assume the score for match is '1', for mismatch, -1 and for the gap (i.e. indel), -2. Thus the similarity score of the sequence can be calculated in the following manner:

Similarity score = (No. of matches × Score for one matche) +

(No. of mismatches × Score of one mismatch) +

(No. of gaps × Score for one gap)

Therefore similarity score between the sequences (X, Y)

$$= (2 \times 1) + \{4 \times (-1)\} + \{1 \times (-2)\}$$
$$= 2 + (-4) + (-2)$$
$$= -4$$

One can consider different alignment form of X and Y, for the proper alignment. For this, gaps can be inserted in the sequence as follows:

```
X = A T T C G T A
    | |   |     |
Y = A T _ C T A A
```

In the above alignment, the number of matches has become 4 with one gap and two mismatches. The alignment or similarity score of the alignment can be calculated as:

$$\text{Similarity score (X, Y)} = (4 \times 1) + \{(2 \times (-1)\} + \{1 \times (-2)\}$$
$$= 4-2-2$$
$$= 0$$

So, it is clear that the similarity or alignment score has been increased after inserting gaps in this alignment.

The following alignment is also possible for the sequences X and Y.

```
X = A T T C G T A
    |   | |     |
Y = A _ T C T A A
```

In this case, alignment score will be 0. Hence, the two sequences can have more than one type of alignment options but in each of the alignment, the alignment score would be same.

In case of very small sequences (as in above example), the alignment can be done manually but in case of large sequences, computational analysis is required. There are several computational programs available over the world wide web. However, each follows a specific algorithm to compute the sequence alignment.

SEQUENCE ALIGNMENT ALGORITHMS

The two main algorithms used for pair wise alignment of the sequences are:

1. Needlemann and Wunsch (Global pairwise sequence alignment)
2. Smith and Watermann (Local pairwise sequence alignment)

1. Needlemann and Wunsch Algorithm

Needlemann and Wunsch (1970) have proposed a very effective methodology of global sequence alignment which is based on the Dynamic Programming Approach. In this method one can consider four cells and their addresses to construct a matrix (Fig. 1).

	i-1	*i*
j-1	***(i-1,j-1)***	***(i,j-1)***
j	***(i-1,j)***	***(i,j)***

Fig. 1. Structure and addresses of the cells in a matrix.

If the second column is ***i***, then the pervious column would be ***i-1***, similarly if the row is ***j*** then its previous row would be ***j-1***. In this way one can assign the addresses to each cell. The cell which is common in the ith column & j^{th} row will have the address ***(i,j)***. Thus it can be observed that the address of any particular cell includes the combined addresses of the rows and columns.

Needlemann and Wunsch algorithm can be easily explained with the help of the following example.

Consider alignment of two sequences X and Y.

X = A T T C G T A
Y = A T C T A A

Let 'm' be the length of sequence X and 'n' is the length of sequence Y.

Therefore:
$m = |X| = 7$
$n = |Y| = 6$

As explained earlier for the alignment of two sequences, gaps are inserted in the sequence at any desired position. According to the algorithm a matrix M of dimensions $(m+1)\ X\ (n+1)$ can be created. In the length of the sequences, 1 will be added because a gap may be introduced in either sequence to obtain the alignment. One can choose any values of the scores of match, mismatch and gap. For this example, let us assume the scores for match = +1, mismatch = –1, gaps = –2. We can define sequence X and Y as.

$X = X_1 \ldots\ldots\ldots\ldots X_i \ldots\ldots\ldots\ldots X_{m.}$
$Y = Y_1 \ldots\ldots\ldots\ldots Y_j \ldots\ldots\ldots\ldots Y_{n.}$

Let $\sigma (X_i, Y_j)$ is the score of the alignment of ith alphabet of sequence **X** with j^{th} alphabet of sequence **Y** (including spaces). If X_i & Y_j are matched then $\sigma (X_i, Y_j)$ = ***match score = +1***, if X_i & Y_j are mismatched then $\sigma (X_{i,} Y_j)$ = mismatch score = –1 and if either X_i or Y_j is aligned with gap i.e. $\sigma (x,-) = \sigma (-, y)$ = ***gap score = –2.***

The matrix M can be drawn with the dimensions $(m+1) \times (n+1)$ *i.e.* $= 8 \times 7$. The first column and row will have zero address (0) and successively $X_1 \ldots\ldots\ldots X_i \ldots\ldots\ldots X_m$ and $Y_1 \ldots\ldots\ldots Y_i \ldots\ldots\ldots Y_m$ for columns and rows, respectively.

Let $V_{(i\,,\,j)}$ is the optimal score of alignment of $X_1 \ldots\ldots\ldots_i$ & $Y_1 \ldots\ldots\ldots_j$.

(Where $0 \leq i \leq m$ ***and*** $0 \leq \sigma j \leq n$ that is i may have values ranging from 0 to m and similarly j may have values ranging from 0 to n)

The base condition can be defined as:

$$V(i,0) = \sum_{k=0}^{i} \sigma(X_k, -) \qquad \ldots\ldots\ldots\ldots(1)$$

$$V(0,j) = \sum_{k=0}^{j} \sigma(-, Y_k) \qquad \ldots\ldots\ldots\ldots(2)$$

CONSTRUCTION OF MATRIX

A binary matrix is constructed from the sequence alignment by placing them on X- and Y- axis in a graph as shown in Fig. 2.

Y \ X	-	A	T	T	C	G	T	A
-	0							
A								
T								
C								
T								
A								
A								

Fig. 2. Empty matrix *M*.

The cell common in first row and first column will be filled with **0** since in this case the zero will be the score of alignment of a gap region. To fill the remaining cells of first row and column we will follow the base condition. We have the value of $V_{(0,0)}$ =**0** (Fig. 2). ***V(i,0)*** is the score of the cell having address ***(i,0)*** which will be equal to sum of the score of previous cell and the alignment score of kth alphabet of sequence X. ***k*** is a variable which can have values from ***0*** to ***i***. In first step ***k = i = 0***, for second step ***k*** will change to 1 as ***k = i = 1.....i.*** In this way the equation (1) will calculate the values for all cells in first row. Similarly, in equation (2), ***k = j = 0,1......j*** and thus it will calculate the values of all cells of first column. Thus, base condition is used to fill the first row and first column of the matrix (Fig. 3)

RECURRENCE RELATIONS

For $1 \leq i \leq m,\ 1 \leq j \leq n$

$$V_{(i,j)} = \max \begin{bmatrix} V_{(i-1,j-1)} + \sigma(X_i, Y_j) & \text{..........} & \text{(I)} \\ V_{(i-1,j)} + \sigma(X_i, -) & \text{..........} & \text{(II)} \\ V_{(i,j-1)} + \sigma(-, Y_j) & \text{..........} & \text{(III)} \end{bmatrix} \quad ..(3)$$

X →

Y ↓

0 + gapscore(-2)=-2

Previous score (-2) + gapscore (-2)= -4

	-	A	T	T	C	G	T	A
-	0	-2	-4	-6	-8	-10	-12	-14
A	-2							
T	-4							
C	-6							
T	-8							
A	-10							
A	-12							

Fig. 3. Matrix M after the base condition.

Recurrence relation is to fill rest of the matrix. To do this, start at 2nd row and 2nd column. Let the 2nd column be 'i' and 2nd row be 'j' as explained in Fig. 4.

To calculate alignment score $V_{(i,j)}$ for the 2nd column and 2nd row, equation (3) can be used which have 3 sub equations I , II and III. From the first sub equation (I), calculate the sum of the value ***(i-1, j-1)*** in the cell and alignment score of X_i with Y_j i.e. $s\ (X_i,\ Y_j)$. Sub equation (II) will calculate the sum of value in cell ***(i-1, j)*** and alignment score of ith alphabet of sequence X with gap. Similarly, sub equation (III) will calculate sum of the value in cell ***(i,j-1)*** and alignment

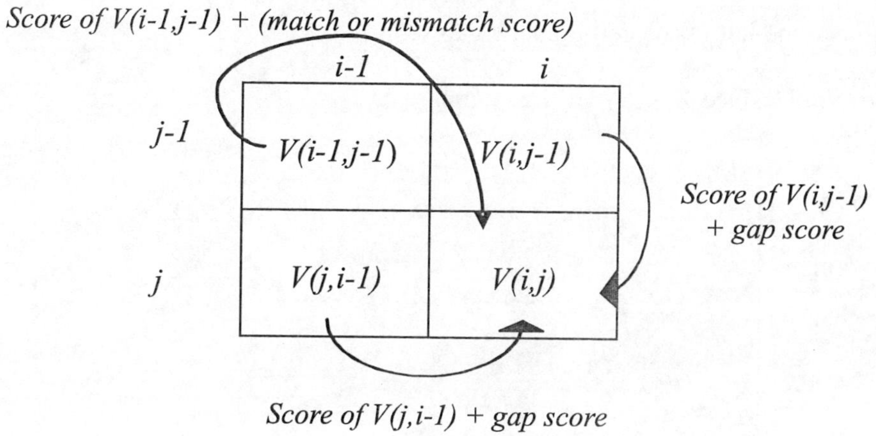

V(i,j) = The maximum of these three values

Fig. 4. Calculation of V(i,j).

of jth alphabet of sequence Y with gap. Now we have three values, one from each sub equation. The maximum of these three will be the $V_{(i,j)}$. (Fig. 4 and 5).

Each cell (Fig. 5) will get maximum of three possible values:

X → (*i*)

Y ↓		-	A	T	T	C	G	T	A
	-	(i-1,j-1) 0	(i,j-1) -2	-4	-6	-8	-10	-12	-14
	A	(i-1,j) -2	(i,j) +1						
(*j*)	T	-4							
	C	-6							
	T	-8							
	A	-10							
	A	-12							

7

Fig. 5. Filling of the matrix.

1. The value to the upper left (diagonal) of the cell plus the match or mismatch score.
2. The value above the cell plus the gap score
3. The value to the left of the cell plus the gap score.

In our example, since $X_i = A = Y_j$, i.e. match = + 1

$$V_{(i,j)} = \max \begin{bmatrix} V_{(i\text{-}1,j\text{-}1)} + \sigma(X_i, Y_j) = 0 + 1 \quad \_\_\_\_ = 1 \\ \small (0) \quad + \text{(match score (+1))} \\ V_{(i\text{-}1,j)} + \sigma(X_i, -) = -2 + (-2) \quad \_\_\_\_ = -4 \\ \small (-2) \quad + \text{(gap score (-2))} \\ V_{(i,j\text{-}1)} + \sigma(-, Y_j) = -2 + (-2) \quad \_\_\_\_ = -4 \\ \small (-2) \quad + \text{(gap score (-2))} \end{bmatrix}$$

$$V_{(i,j)} = \max \begin{bmatrix} +1 \\ -4 \\ -4 \end{bmatrix} = +1 \text{ (maximum score of } V_{(i,j)} = +1)$$

After calculating $V_{(i,j)}$., we can move one column towards right side and calculate the values in other cells as shown in Fig. 6.

X → (*i*), Y ↓ (*j*)

	-	A	T	T	C	G	T	A
-	0	-2	-4	-6	-8	-10	-12	-14
A	-2	+1	-1					
T	-4							
C	-6							
T	-8							
A	-10							
A	-12							

Fig. 6. Moving towards right hand side while filling the matrix.

Now consider the next set of 4 cells and calculate the alignment score of *T* with *A*. Follow the same recurrence relation to find the values. Calculate values from 3 sub equations and choose the maximum value. In this way, fill all remaining cells one by one. (Fig. 7).

The value in the last cell is known as **Optimal Score** or **Alignment Score**.

X → (*i*), Y ↓ (*j*)

	-	A	T	T	C	G	T	A
-	0	-2	-4	-6	-8	-10	-12	-14
A	-2	+1	-1	-3	-5	-7	-9	-11
T	-4	-1	2	0	-2	-4	-6	-8
C	-6	-3	0	1	1	-1	-3	-5
T	-8	-5	-2	1	0	0	0	-2
A	-10	-7	-4	-1	0	-1	-1	+1
A	-12	-9	-6	-3	-2	-1	-2	0

Optimal or alignment score (the last cell, 0)

Fig. 7 Completely filled matrix M.

TRACE BACK

After filling matrix ***'M'*** completely, the main objective is to obtain the alignment from this matrix. For this purpose, one has to trace back through the matrix. Tracing back is the process of finding the best possible paths of the alignment. To find this, start from the optimal score to find out the values from where the optimal score has been calculated. For this, one has to recall the recurrence relations to find out the cell from where the optimal score has come. In this example the optimal score has come from -1 (since A of Seq. 'X' is matching with A of 'Y'). From the equations of recurrence relation, these are the three values: **-1** (from 1 such that 1+ (-2) = -1)., **0** (from -1 such that : -1+1=0) and **-4** (from horizontal -2 such that: -2+ (-2)= -4). It can be observed that 0 is the maximum and optimal and has been calculated from -1 (diagonal cell). Thus move back towards diagonally placed -1. Now consider -1 and trace back in the same manner until one reaches to zero of first cell (Fig. 8). While tracing back, if one gets two points then trace back the points independently up to zero of first cell. At each bifurcation, there will be two paths. In some rare cases, three paths can also be obtained.

From the figures, two paths can be considered to obtain alignment.

How to write aligned sequence?

To write aligned sequences, first write any of the two sequences. For example, first write sequence ***X*** and then write sequence ***Y*** against sequence ***X*** in an aligned form. Now start writing ***Y*** from the right side and put one alphabet for each diagonal run. In the example, 4

X →

Y ↓	-	A	T	T	C	G	T	A
-	0	-2	-4	-6	-8	-10	-12	-14
A	-2	+1	-1	-3	-5	-7	-9	-11
T	-4	-1	2	0	-2	-4	-6	-8
C	-6	-3	0	1	1	-1	-3	-5
T	-8	-5	-2	1	0	0	0	-2
A	-10	-7	-4	-1	0	-1	-1	+1
A	-12	-9	-6	-3	-2	-1	-2	0

Fig. 8. Paths of tracing back.

continuous diagonal runs were obtained from the optimal score as follows:

$$\mathbf{0 \rightarrow -1 \rightarrow 0 \rightarrow +1 \rightarrow 0}$$

While filling the matrix, we consider 4 cells and compare two alphabets in which match or mismatch scores are calculated from diagonal cell. Vertical and horizontal cells are used for comparing an alphabet with gap. Since, there are four diagonal runs, put 4 alphabets of ***Y*** as such against sequence *X* as given below:

```
X = A T T C G T A
Y = .  .  . C T A A
```

See in the trace back scheme (Fig. 8), after 4 continuous diagonal runs, there is a bifurcation in the path, in which one is running diagonally and other one is in left direction (horizontal). It is giving two paths to reach up to end.

Consider first path (black), in which fifth run is diagonal, it means one more alphabet of sequence ***Y*** should be put against sequence ***X***. After it, the path is moving toward left. One important thing should be kept in mind while moving either in vertical or in horizontal direction, insert a gap in the sequence (towards the direction of movement) before the alphabet of the corresponding column or row. In the example, while moving along path I (black), there is one horizontal movement towards sequence *Y*. The alphabet is '***A***' of the corresponding row. Thus, insert a gap before '***A***' in sequence *Y*. After horizontal movement, there is again a diagonal movement, so put another alphabet of sequence ***Y*** that is '***A***'. In this way, the complete alignment can be obtained according to path I.

```
X = A T T C G T A
    |   | |     |
Y = A _ T C T A A
```

One can obtain another alignment by tracking the 2nd path (red) as:

```
X = A T T C G T A
    | |     | |
Y = A T _ C T A A
```

The alignment can be validated by calculating the alignment scores as follows:

$$\begin{aligned}\text{Alignment score} &= (4 \times \text{match}) + (2 \times \text{mismatch}) + (1 \times \text{indel}) \\ &= (4 \times 1) + (2 \times (-1)) + ((-2) \times 1) \\ &= 4 + (-2) - 2 = 4 - 4 = \mathbf{0} \text{ (optimal score)}\end{aligned}$$

Smith and Watermann algorithm

In many applications two sequences may not be highly similar, but may contain subsequences with high resemblance. Therefore, it is important to find, extract and align a pair of subsequences that possess the highest similarity. This is known as local alignment problem. Local alignment problem can be solved through Smith and Watermann algorithm which is also based on dynamic programming. To explain this, the same example of Needlemann and Wunsch algorithm can be taken here:

Base condition

In Smith and Watermann algorithm, the base condition is quite different from the Needlemann and Wunsch algorithm. In this case, the value in all cells of first row and column is limited as zero i.e. all cells of first row and first column of the matrix will have value equal to **0.** (Fig. 9)

$$V_{(i,0)} = 0 \qquad \text{.....................} \quad (4)$$

$$V_{(0,j)} = 0 \qquad \text{.....................} \quad (5)$$

Recurrence relation

The recurrence relation is very much similar to Needlemann and Wunsch algorithm with a small difference in the calculations of $V_{(i,j)}$, where we choose the maximum of 4 values.

X →

Y ↓

	-	A	T	T	C	G	T	A
-	0	0	0	0	0	0	0	0
A	0							
T	0							
C	0							
T	0							
A	0							
A	0							

Fig. 9. First row and column is filled by base condition in Smith and Watermann algorithm.

$$V_{(i,j)} = max \begin{bmatrix} 0 \\ V_{(i-1,j-1)} + \sigma(X_{I,} Y_j) \\ V_{(i-1,j)} + \sigma(X_{i,} -) \\ V_{(i,j-1)} + \sigma(-, Y_j) \end{bmatrix} \qquad \text{.......................} (6)$$

The above relation reveals that if any negative value is obtained then put zero at that place since zero will always be the greatest among negative values.

The value in the last cell is not the optimal score. The alignment score would be maximum of all values in the matrix.

Optimal score = MAXIMUM $\boldsymbol{1 \leq i \leq m, 1 \leq j \leq n\ [V_{(i,j)}]}$ .... (7)

Let us consider the same example:

X = A T T C G T A
Y = A T C T A A

Create the matrix similar to Needleman and Wunsch of **$(m + 1) \times (n + 1)$** dimension with values as match = 1, mismatch = -1, gap = –2 as shown in Fig. 10.

According to equation (7), one can observe that 2 is the maximum value in the entire matrix. Thus, the optimal score will be equal to **2**.

Y ↓ \ X →	-	A	T	T	C	G	T	A
-	0	0	0	0	0	0	0	0
A	0	1	0	0	0	0	0	1
T	0	0	(2)	1	0	0	1	0
C	0	0	0	1	(2)	0	0	0
T	0	0	1	1	0	1	1	0
A	0	1	0	0	0	0	0	(2)
A	0	1	0	0	0	0	0	1

Optimal or alignment score

Fig. 10. Matrix filled by Smith and Watermann algorithm showing the optimal scores in the matrix.

Trace back

Like Needleman and Wunsch algorithm, one can trace back from the optimal score of the alignment. Unlike Needleman and Wunsch algorithm, one can get the optimal score anywhere in the matrix that could be repeated more than one time. In the above example, the optimal score is **2** at three places. One can start tracing back the alignment from all three places and will move until one gets the first zero in the path (Fig. 11).

From the figure 11, we can observe that we have 3 possible aligned substrings of sequences ***X*** & *Y*.

Substring 1

X = A T T C G | T A |
Y = ——————— | T A |

This alignment has been obtained from the **BLACK** path in which, there are only two diagonal runs comparing two alphabets of sequence X with two alphabets of sequence Y.

Substring 2

X = A T | T C | G T A
Y = —— | T C | ————

X →

Y ↓

	-	A	T	T	C	G	T	A
-	0	0	0	0	0	0	0	0
A	0	1	0	0	0	0	0	1
T	0	0	2	1	0	0	1	0
C	0	0	0	1	2	0	0	0
T	0	0	1	1	0	0	1	0
A	0	0	0	0	0	0	0	2
A	0	0	0	0	0	0	0	1

Fig. 11. Matrix M showing three possible paths of alignment.

This alignment has been obtained from the **BLUE** path in which, there are only two diagonal runs comparing two alphabets of sequence X with two alphabets of sequence Y.

Substring 3

X = | A T | ______________

Y = | A T | ______________

This alignment has been obtained from the **RED** path in which, there are only two diagonal runs comparing two alphabets of sequence X with two alphabets of sequence Y.

All alignments would be accurate because they are at different regions of the sequence.

Basic difference between Needlemann Wunsch & Watermann & Smith algorithms.

Needleman & Wunsch algorithm	Waterman & Smith algorithm
1. Perform alignment globally.	1. Perform alignment locally.
2. Score may have negative value.	2. Score cannot be negative.
3. Last value of the matrix is optimal score.	3. Highest value of the matrix is optimal score.
4. Important for comparing two sequences.	4. Important for comparing substrings of the two sequences.

Suggested Readings

Baxevanis A.D. and B.F.F Ouellette (2001), Bioinformatics: A practical guide to the analysis of genes and proteins. John Wiley & Sons, Inc. NY, USA.

Brown S.M. (2000), Bioinformatics: A biologists guide to computing and the internet. A Biotechniques Books publication Eaton Publishing.

Haubold, B. and T. Wiehe (2006), Introduction to computational biology: An evolutionary approach. Birkhauser Verlag, Basel-Boston-Berlin, Germany.

Needleman, S.B. and C.D. Wunsch (1970), A general method applicable to the search for similarities in the amino acid sequence of two proteins. *J. Mol Biol.* **48**: 443-453.

Smith, T.F. and M.S. Waterman (1981), Identification of common molecular subsequences. *J. Mol. Biol.* **147**: 195-197.

6

Similarity Searches Software and their Applications

In the previous chapter we have seen the basic concepts and algorithm used in sequence alignment. Sequence alignment has very important application in finding similarity between two or more organisms at DNA level. Once a new DNA sequence is found by the scientists from any biological sample, they would be always curious to know what that sequence contains. Does it have some gene(s)? What trait that gene controls? How many organisms have such similar genes? Hence, one can ask so many questions and try to find out their answers by performing sequence similarity searches against publically available genome sequence databases.

SEQUENCE SIMILARITY

Sequence similarity and sequence homology are two important terms which have different meanings. Sequence homology refers to the similarity due to descent from common ancestor which means having common function or 3 D structures of the proteins. Sometimes homology can be inferred from the similarity. For instance, 25% similarity in a string of 100 amino acids is a strong evidence of homology. Similarity can be measured in two different forms:

i) **Quantitative terms**, which represent degree of similarity e.g. similarity score indicating 40% match between two DNA sequences,

ii) **Qualitative terms**, which is basically an alignment which shows the regions of two sequences which are either similar or different. The most correspondences and the least differences between two sequences is called the optimal alignment. One can see following two sequences and look for the similarity between these.

a) TCGATCGGCA

b) CTGATCTTAAGC

There are many ways of performing similarity between these two sequences. One can align these from one end to another and see how many of the bases are matching to each other. One can also align them as follows:

```
T C - G A T C G T C - A
- C T G A T C T - C G A G C
g . g . . . . m g . g .
```

In above alignment **g** means gap, **dot** means exact match and **m** means mismatch.

The alignment can be of two types i.e. local and global alignments. The details of these two types of alignments and their basic algorithms have already been explained in the previous chapter. The commonly used algorithm is Smith-Waterman's in both Basic Local Alignment Search Tool (BLAST) and FASTA softwares. As explained earlier, Smith-Waterman algorithm is a rigorous approach based on dynamic programming and is very robust in finding local alignments between a query sequence and the database of similar sequences. In this chapter we will discuss various tools which can be used for performing sequence (DNA/protein) alignments.

Amino Acid Substitution Matrices

During the course of evolution some amino acids substitute for others in the related proteins based on their physio-chemical properties. The physico-chemical properties like hydrophobicity, polarity, size of the side chain, negative and positive charges, and aliphatic vs non-aliphatic distinguishes amino acids from each other. Beside these, due to the degeneracy of the amino acid codon, the mutational differences vary from one to three steps in different amino acids. Because of the differences in the physico-chemical properties and mutational differences the sequence alignment score is given. For instance, small hydrophobic amino acid like valine can easily be substituted for isoleucine. Therefore, while comparing two protein sequences, the alignment score should be given more to the identical amino acids compared to the substituted one. Similarly conservative amino acids should also get higher alignment score compared to the non conservatives while comparing similar protein sequences. These scores are mainly based on log-odds scoring matrices i.e. each score in the matrix is the logarithm of an odds ratio. In those cases where amino acid residues are randomly replaced, the odds ratio is the ratio of the number of times amino acid residue "X" is observed to replace residue "Y" divided by the number of times residue "X" would be expected to replace residue "Y". In different substitution matrices, the positive substitution score shows that the pair of residues replace each other more often than expected by chance hence the aligned sequences would be homologous, whereas the negative score indicated that the pairs of residues replace each other less than would be expected by chance, thus the aligned sequences would not be homologous. Two types of substitution matrices have been briefly explained below. For more details readers can refer to the suggested readings.

1. Point Accepted Mutation Matrices

The Point Accepted Mutation (PAM) Matrices, one of the first scoring matrices being used extensively developed by Dayhoff and her associates in 1970 (Dayhoff *et al.*, 1970). They constructed a phylogenetic tree for each family of the protein by aligning all the proteins belonging to several families and substitution found on each branch of the tree was recorded. They developed a frequency table in which the rates of substitution of amino acids for each other are given based on their replacement over a short evolutionary period. The numbers given in the PAM matrices table indicate the number of mutation per 100 amino acids for that particular gene. According to this scoring system, one PAM is an evolutionary divergence unit of which 1% of the amino acids have the tendency to change position upto a certain limit in different organisms. PAM250 matrix means occurrence of 250 point mutations per 100 amino acids in the gene. Different PAM matrices have been developed and are used based on the type of sequences used for making comparisons. For instance, if only one matrix has to be used, then PAM120 would be the most useful. However, for more comprehensive results, more than one matrice like PAM40, PAM120, PAM250 etc. should be used. Even two matrices like PAM80 and PAM200 can also give good coverage while comparing the sequences. Best alignment can be obtained from a diverged pair of sequences corresponding to a specific PAM matrix.

Advantages of PAM matrices

These matrices are better over other methods because:

1. These are based on the process of creating observed mutations.
2. The criteria for selection and fixing of a mutation within a population is used in PAM.
3. It is based on most accurate changes in amino acids compositions which are expected after a given number of mutations in a population.

2. Blocks Substitution Matrices

The Blocks Substitution Matrices which is popularly known as BLOSUM has been developed by Henikoff and Henikoff (1992). These matrices are based on the BLOCKS of ungapped protein alignments. It is basically an improvement over PAM due to the following factors: i) When BLOSUM matrices were developed many sequences of proteins were known in the databases which were used for many more amino acid substitution analysis. ii) The BLOSUM matrices are constructed based on observed substitution within the conserved blocks in multiple sequence alignments. Basically the sequence data is passed through three different stages while constructing BLOSUM matrix.

i. BLOSUM matrix is built by eliminating sequences which are identical in more than x% of their amino acid sequence. The sequences are either removed from the blocks or replaced with the sequences of similar clusters. Thus the matrix constructed from the blocks with less than x% of similarity is called BLOSUM-x . For instance, the matrix which is built by using not more than 45 of the sequences is called BLOSUM45.

ii. In the second stage, the pairs of amino acids in each column, of multiple alignment are counted and their probability of substitution is calculated.

iii. During third stage, the log odd ratio is calculated and entered for each pair of amino acids in the BLOSUM -x matrix. BLOSUM62 matrix is one of the most commonly used BLOSUM matrices.

Advantages of BLOSUM matrices

1. The substitution is based on well conserved blocks representing most reliable alignments hence reducing the proportions of false substitutions.
2. The conserved block regions are mostly found in the databases. Therefore, most appropriate substitution patterns are represented in the matrices upon database searches.

Comparison between PAM and BLOSUM matrices

PAM Matrices	**BLOSUM Matrices**
PAM is based on an evolutionary model using phylogenetic trees.	BLOSUM assumes no evolutionary model, but rather conserved "blocks" of proteins are used here.
It is designed to track evolutionary relationship among the sequence.	It is designed to find conserved domains among the sequences.

PAM and BLOSUM matrices equivalence

PAM	BLOSUM
PAM250	BLOSUM45
PAM200	BLOSUM52
PAM160	BLOSUM60
PAM120	BLOSUM80
PAM100	BLOSUM90

Selection of Matrices for Different Types of Alignments

Short alignment with high similarity	General alignment	Distant sequences
PAM40 BLOSUM90	PAM120 BLOSUM62	PAM250 BLOSUM30

The selection of appropriate matrix is based on the specific sequence alignment. However, BLOSUM62 is used as a default setting in BLAST programme and can be changed from the drop box. For more details, one can go through the suggested readings.

Nucleotide and Protein Codes Supported by the Softwares

The nucleotide and amino acids codes supported by similarity searches software like BLAST and FASTA are given in Boxes 1 and 2.

A	:	adenosine	M	:	A C (amino)
C	:	cytidine	S	:	G C (strong)
G	:	guanine	W	:	A T (weak)
T	:	thymidine	B	:	G T C
U	:	uridine	D	:	G A T
R	:	G A (purine)	H	:	A C T
Y	:	T C (pyrimidine)	V	:	G C A
K	:	G T (keto)	N	:	A G C T (any)
– gap of indeterminate length					

Box 1. Nucleotide codes

A	alanine	P	proline
B	aspartate or asparagine	Q	glutamine
C	cystine	R	arginine
D	aspartate	S	serine
E	glutamate	T	threonine
F	phenylalanine	U	selenocysteine
G	glycine	V	valine
H	histidine	W	tryptophan
I	isoleucine	Y	tyrosine
K	lysine	Z	glutamate or glutamine
L	leucine	X	any
M	methionine	*	translation stop
N	asparagine	–	gap of indeterminate length

Box 2. Amino acids codes

BLAST

Basic Local Alignment Search Tool (BLAST) is one of the most popular software (Altchul *et al.*1990). for performing database search by using either nucleic acids or protein sequences as a query. It can be freely obtained from the National Centre for Biotechnology Information (NCBI) to search GenBank Database on the world wide web (http://www.ncbi.nlm.nih.gov/BLAST). However, the software can also be freely downloaded from the NCBI and customized on laboratory servers for performing local BLAST.

How BLAST Works?

Once a query sequence is used for the BLAST search, it performs a few basic steps and systematically the sequence is searched against the selected databases. Different steps used by this software for performing database search are given in the flow diagram shown in Box 3.

The query sequence which is a long string of either nucleotide or amino acids is first broken into small pieces called "words". As a default setting the DNA sequences is broken into 11 consecutive letters (world length) and amino acids into 3 letters. However, users can change 'world length' as desired.

↓

All the words that are similar both in query and database sequences are identified based on a predefined threshold scores.

↓

The occurrences of the query words are found in the database based on the basic principle of local alignment using Smith-Watermen algorithm. The words (two or more) which appear consistently in the same position are joined to each other.

↓

The alignments are extended to both left and right directions. The pair of sequences which shows local alignments is called High-scoring segment pairs (HSPs)

↓

The score and statistics of the alignments are calculated and the results are depicted on the output window.

↓

The repeat sequences and low complexity regions in the query sequences are masked with the filters provided as a default setting in the BLAST.

BOX 3. Different steps used by BLAST software to perform sequence searches.

In summary BLAST search can be seen as three-step procedure in which

i) A list of high scoring word is first compiled.
ii) It searches for the hits, which is also called seeds.
iii) The seeds are extended in both left and right directions.

Different BLAST Options

Based on the type of query sequence whether it is protein or DNA, a specific program is selected for performing blast search. Different BLAST options and their descriptions are given in Table 1.

Table 1. BLAST options, types of query and database sequence.

Programme*	Compare a query sequence	Against database
Blastn	Nucleotide sequence	Nucleotide sequence
Blastp	Amino acid	Protein sequence
Blastx	A nucleotide sequence translated in all reading frames	Protein sequence
Tblastn	Protein Sequence	A nucleotide sequence database dynamically translated in all reading frames
Tblastx	The six-frame translations of a nucleotide sequence	Six-frame translations of a nucleotide sequence database

*The BLAST search pages allow you to select from several different programs.

Besidee above programs, other specific BLAST options have also been provided in the NCBI. These are Gapped BLAST and Position Specific Iterated-BLAST (PSI-BLAST). The Gapped BLAST is used when we want to add more gaps in the alignment so that longer continuous alignments can be produced. It results in the output of biologically more significant hits as compared to ungapped BLAST. Due to addition of deletion and insertion of gaps in the alignment the entire process has become very fast generating long alignments, thus increasing the speed of the prgrams in performing searches against the ever growing size of the biological databases.

PSI- BLAST

Position Specific Iterative – BLAST (PSI-BLAST) is based on the principle that the conserved patterns of the alignments of related sequences may help in identification of distant similarity among the sequences. These patterns are named position-specific score matrices, and Hidden Markov Models, motifs or profiles. A specific score is assigned to each amino acid present at the specific position of the derived patterns. For instance, the highly conserved residue at a specific position is assigned a high positive score and other residues are given as high negative scores. At weakly conserved position' residues are assigned zero or nearly zero score. The potential insertion or deletions are assigned position specific scores. The iteration of search also enhanced the power of profile method. Different steps used in PSI-BLAST are given in Box. 4.

A single protein sequence is used as an Input in this program

↓

Comparison to a protein database, using the gapped BLAST program

↓

From the significant local alignments a multiple alignment is constructed

↓

The template for the multiple alignment and profile is the original query

↓

At different template positions different number of sequences aligned

↓

Using local BLAST, the profile is compared to the protein database

↓

The statistical significance of the local alignments is estimated

↓

At the final step, the PSI-BLAST iterates a specified number of times or until convergence is obtained by returning to step (2)

Box 4. Different steps used by PSI-BLAST

Once a query sequence is pasted in the BLAST search window, next step is to select the appropriate database against which the search is performed. Various databases are available in the NCBI of which some are explained below.

Database Selections

For performing BLAST, one of the important steps is to select appropriate database against which the query sequence has to be searched for possible matches. One can make species specific BLAST search by selecting the data base from a drop down box. Different types of database are available in the GenBank. Some of the examples of protein and nucleotide sequences are given in Table 2.

FASTA

Another important alignment programme called FASTA (pronounced FAST-AYE and stands for **FAST**-**A**LL) was developed by Dr. William Parson in 1997. It also works on the basic concept of local alignment with little variation. In FASTA, database search is accelerated by using several passes of query sequence over the database. Then it retains a best match of a subset for further analysis. In the first pass, it detects a stretch of short sequence known as 'words' which shows similarity with the database. The small word size of the FASTA is called *k-tuple* or KTup. The default setting of *k-tuple* is 6 for nucleic acids and 2 for protein i.e. the software first break

Table 2. Different types of protein and nucleotides databases available for performing BLAST search.

Protein databases	
Database	**Description**
nr	Non-redundant GenBank CDS translations + PDB + SwissProt + PIR + PRF, excluding those in env_nr.
refseq	Protein sequence from NCBI reference project.
swissprot	Last major release of the SWISS-PROT protein sequence database (no incremental updates).
pat	Protein from the Patent division of GenBank.
month	All new or revised GenBank CDS translations + PCB + SwissProt + PIR+ PRF released in the last 30 days.
pdb	Sequence derived from the 3-dimensional structure records from the Protein Data Bank.
env_nr	Non-redundant CDS translation from env_nt entries.
Nucleotide databases	
nr	All GenBank+ EMBL+DDBJ+PDB sequences (but no EST, STS,GSS, or phase 0,1 or 2 HTGS sequences). No longer "non-redundant" due to computational cost.
refseq_mrna	mRNA sequences from NCBI Reference Sequence Project.
refseq_genomic	Genomic sequences from NCBI Reference Sequence Project
est	Database of GenBank+EMBL+DDBJ sequences from EST division.
est_human	Human subset of est.
est_mouse	Mouse subset of est.
est_others	Subset of est other than human or mouse.
gss	Genome Survey Sequence includes single-pass genomic data, exon trapped sequences, and Alu PCR sequences.
htgs	Unfinished High Throughput Genomic Sequences: phase 0, 1 and 2. Finished, phase 3 HTG sequences are in nr.
pat	Nucleotides from the Patent division of GenBank.
pdb	Sequences derived from the 3-dimensional structure records from Protein Data Bank. They are NOT the coding sequences for the corresponding proteins found in the same PDB record.
month 30 days.	All new or revised GenBank+ EMBL+DDBJ+PDB sequences released in the last
alu_repeats	Select Alu repeats from REPBASE, suitable for masking Alu repeats from query sequences.
dbsts	Database of Sequence Tag Site Entries from the STS division of GenBank+EMBL+DDBJ
chromosome	Complete genomes and complete chromosomes from the NCBI References Sequences project. It overlaps with refseq_genomic.
wgs	Assemblies of **W**hole **G**enome **S**hotgun sequences.
env_nt	Sequence from environmental samples, such as uncultured bacterial samples isolated from soil or marine samples. This does overlap with nucleotide nr.

nucleotide sequence in a word size of 6 and then search its similarity in the database. A perfect match of the words is required in case of FASTA hence weak but significant similarity between protein sequences is often ignored.The short region of similarity then extends into alignment with gaps. For the calculation of score of alignment a scoring matrix is used. The best score of these alignments known as **init1** score is then retained. Subsequently, another alignment score known as **initn** is calculated for whole joined region of the sequences. In the final step, sequence regions with high **initn** score are aligned with the query sequence by using a more sensitive method. In the final search, numbers of sequences to be retained in the database for further alignment are decided based on threshold parameters and an optimum score known as **opt** score is calculated. The statistical significance of the alignment is calculated (**Z-score**) on a statistical model which assumes that the alignments are without gaps. The normalized Z -score is basically derived from the opt score and is then converted into *E-value* which tells the probability of a given match between two sequences.

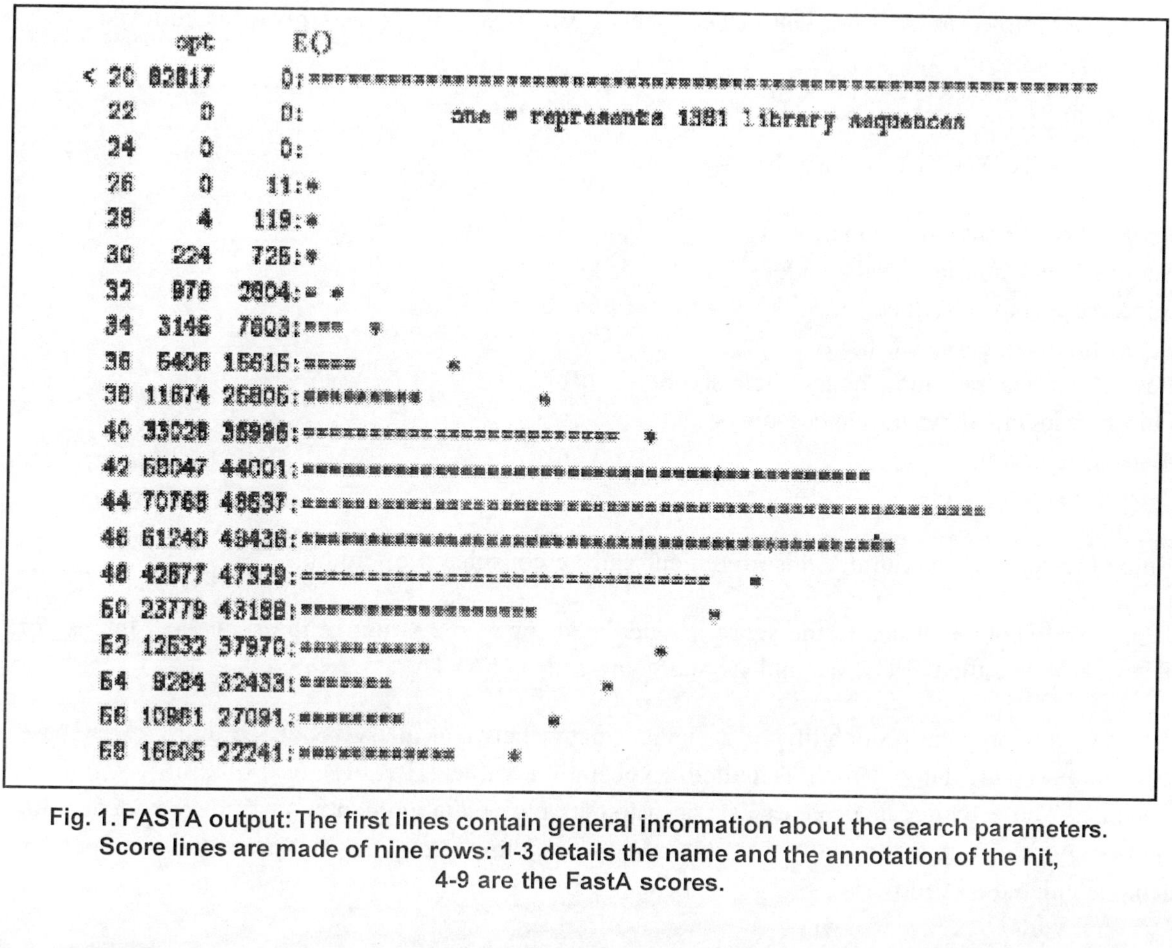

Fig. 1. FASTA output: The first lines contain general information about the search parameters. Score lines are made of nine rows: 1-3 details the name and the annotation of the hit, 4-9 are the FastA scores.

FASTA Output

The FASTA output is presented in a typical histogram form as shown in Fig.1. The histogram shows the distribution of z-score and statistical expectations of the database. The X axis of the histogram shows the z-score printed in the left column and increases from top to bottom. The numbers of sequences matching with the database records having the score are shown in the Y axis. The observed distribution score which is plotted as "=" signs is given in the second column. The third column shows the expected random distribution of the score and is plotted as "*" signs.

Alignment Score and Expectation Values

As discussed in previous chapter the sequence alignment score is calculated based on Smith and Watermen algorithm in case of local BLAST. For this a scoring matrix is made and a different weightage is given based on match, mismatch and gaps in two aligned sequences. Let us take a simple case of two DNA sequences a &b in which score is given as follows:

```
TC-GATCGTC- A - - (a)
  |  | | |       | |
-CTGATCT- CGAGC (b)
```

Score of two identical residues = +1
Score of two non-identical residues = – 1
Score of a gap introduced in either of the sequence = – 1
So in this sequence, we have
Seven identical residues, hence their score = 7
One non-identical residue and score = – 1
4 gaps, hence the score = – 4
Therefore, total score = 7 – 1– 4 = 2

Since the score is maximum, this alignment can be considered optimum.

In this way, computationally the score is calculated for whole string of the sequence. In case of BLAST, it is called Bit score and *Z*- score in case of FASTA.

Besides, a score, the probability of a chance match between query sequence and the database sequence is calculated which is called Expectation value (E-value) or Probability value or P-value. The E-value is used to assess whether a given alignment constitutes evidence for homology. It also helps to know from a chance alone, how strong an alignment can be expected. It is calculated as follows:

e.g. If two sequences of *a* and *b* length are aligned then finding at least one segment pair with score ≥S

Pr (atleast one high scoring segment pairs (HSP) with score $\geq S$) = $1 - \exp\{K\,ab\,e^{-\lambda S}\}$

Where *K* and l depend upon the scoring scheme. This probability is called *P*- value.

However the expected number of segment pairs having score $\geq S$ in the random model is called *E- value*.

i.e. E [number of HSPs with score $\geq S = Kab\ e^{-\lambda S}$]

To get rid of the dependence on scoring system this score is normalized by using following equation

$$S' = \lambda S - Log\ K / Log\ 2$$

Thus the E value $\approx ab/\ 2^{S'}$

In BLAST output we get the normalized score and the *E*-value where *a* is the length of query sequence and *b* is the length of all the sequences present in the database.

While interpreting the output results of BLAST or FASTA there is direct relationship between Bit socre (z-score in case of FASTA) and the *E*-value. i.e higher the bit score and lower the *E*-value between two sequences means the sequences are very closely related to each other. When *E*-value = 0 or near to zero it means that the unknown sequence has perfect match in the database.

Comparison between BLAST and FASTA

BLAST	FASTA
It is very fast. A given sequence may take 2-3 minutes for performing database search.	It is relatively slow and a given sequence may take several hours for database search.
It is more sensitive for detecting protein similarity at default settings.	FASTA is less sensitive for detecting protein similarity at default settings.
It does not require perfect sequence match.	It requires perfect sequence match at the initial stage of searching.
It usually performs erroneous results in the microsatellite regions of the DNA.	It gives better results in microsatellite regions of the DNA because of small KTup.
It may fail together in detecting longest exons because it creates ungapped alignment.	It is better suited for specialized task like detecting genomic regions using cDNA sequences.

Running a BLAST

BLAST search can be easily performed on NCBI website from the web address: http://www.ncbi.nlm.nih.gov/. Various BLAST options are available on this home page. One can even perform BLAST against the full genome sequences of Human, Mouse, Rat, *Arabidopsis thaliana* and *Oryza sativa* etc. Besides, one can also perform specialized BLAST against different databases. For performing BLAST search, the query sequence should be in FASTA format i.e. the first line of the sequence should start from > sign followed by the description of the sequence. In the second row the sequence either DNA or protein should start without any break as shown in Fig. S1. An example of performing BLASTn has been explained with different steps as below:

```
>K-60
CTTAATTGGTCTTCCCTTAAATCGCTGAACTAATGTAAAACTAATCTTCCATGACGACTATTC
GTTTGGCCTCGTTGGAGGGTGGGGTTTTTCACACAATAATACGTCAGAAGTGTTGAAAGAA
ACAACTTCATATCGTGCCAATTCAACTCAAGAGGATATGTATTAATAACTGAACATGAGTTCA
TTCACTTGAACATCAAAAAGAAATAGCAAGGACAACTAATTGTTTGTTTAATCTATCAAGGC
TAGTTCAATTGCTTTAAGAATAGCTCAATGAAAATCAGAAACATGGCTTTCCATGAACAGAG
CACTGATGACATACCTGATGGTTCTTTAAAATTGGGGCAATCTTCAACAATAATGTCGCTACA
TGTTAGGCCTTCAGGAATGGAGTGCAGAATTGGACAGCCAGACAAACACATTATCTGTAGA
AAAGGTAGGTTCTCCAACCATTCTGGCAAAATCTCCAATTTGTTGCAGTTATCAATTTTCAA
GCTGGTGAGAGAGGAAAGATGCTGCAAACCCTGTGGGAGGCATGATAATTCATGACAACC
ACATATCTCCAACTTCTTGAGGCGATCAAGAGCCTGTAGACCCTCACAGCCTGAAGAAGCA
TGCAACTCTTCACAGAAAGAAATTTGAAAGATTTTCGAGGCTCTTCTCTACACCTCTCATAC
CTTCCTTCGAGCTGTACAACATCTTGTTATTCATCCACAAAATCAACTTTTCAATCGATGGAG
ACTCCATATGCAGTGCTCTCAATTTTGGGCACTGAATGATGACCAGTTCTGCAAGACGAGG
AAACAGATTCATCCTAGATGACTGTTCCCAGACTTCAAGGCTTGTCATGTATGATAGAATCA
GCTTCTCCAGTGAAGGGAATGTGCCGTTATGACCATAGAAATTATCACCAATAGAACTGAGA
TTATCAACTCCACTTATCTCTGCAATCTTCAGACATGGTAGTATCCCAAGTGGAGGTAATGCC
TTTTCACATGCTCTCATATTAACTAGTCTTATCTCAACCAGAGATGTAAGATATGGCTCGGTT
CTTGTCATCCAAGATGGAAATACATATCCTTCATACGCAACAATCTCCAAAGTTTTCAGGCA
CTGGCTAGGTTGGAGGGTTTCAAGTACTTCATGATCTATTCTACTGGCATTTCCAGCATCCAT
ATTCCACCTCGACATCAATGTCTCGAGCTTTTCTTTGTCTTGAAGTTTTGCCATTCTTGCATC
TTCAGTGTCTGAAACCTTTTCAAGGTTCACTAGTGATAGTCTATTGAGGTCTGGCAGTGATT
GAAGCTCTGACATCGCACTACTTCCATTGTTGGGGACAAAGTAACCAAGCAATGTGTGAAG
ATTTGTAAGTTGGCCCATCGCAGAGGCATTGCTGTGAGTGAGTAGCAATCGAGAACATTAGT
ACTGAGC
```

Fig. S1. DNA sequence in FastA format used for BLAST search.

Stept 1. Go to home page of the NCBI and CLICK on Blast (Fig. S2)

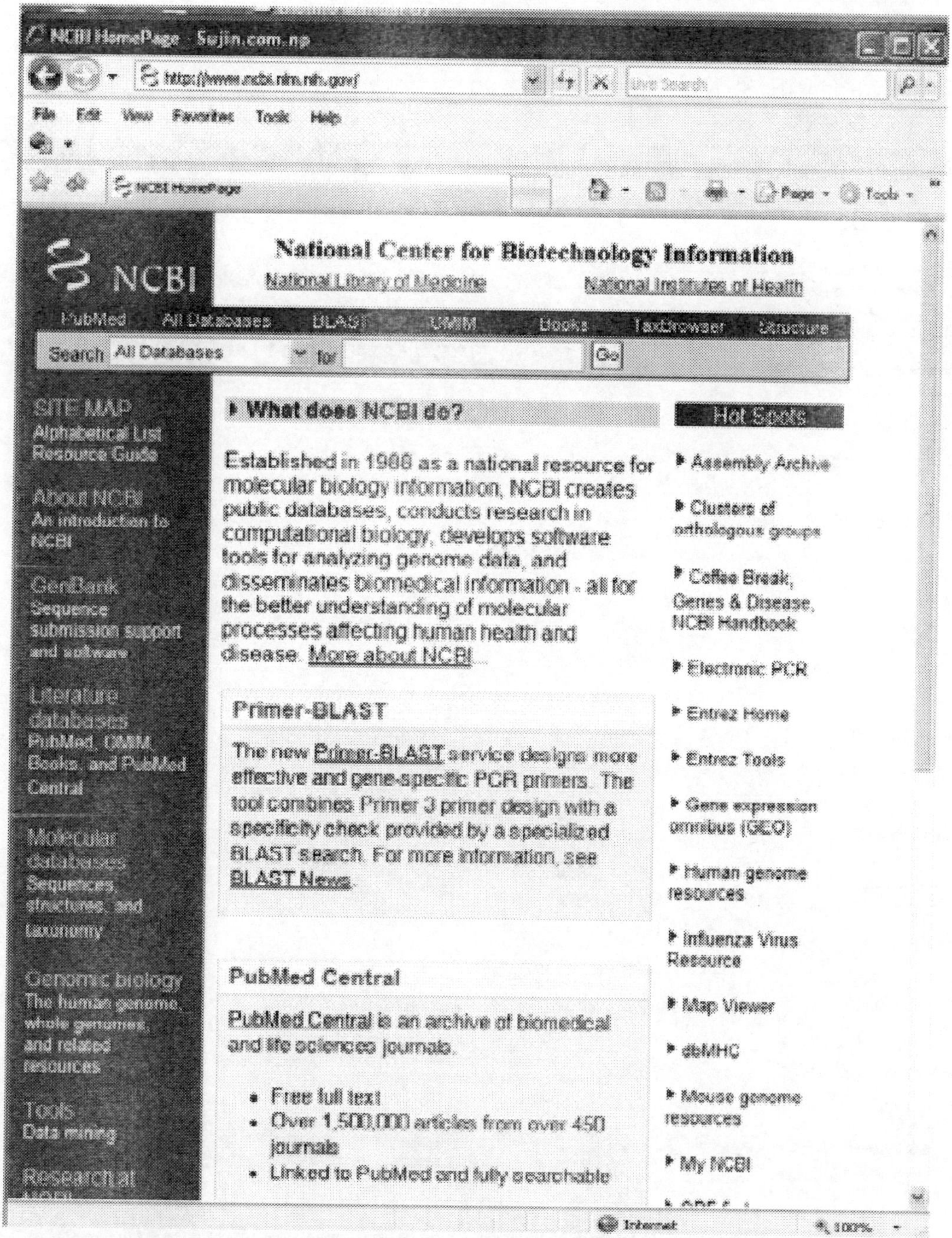

Fig. S2. NCBI home page.

Stept 2. Select type of BLAST Options from this window. For instance for nucleotide-nucleotide BLAST CLICK on BLASTN (Fig. S3)

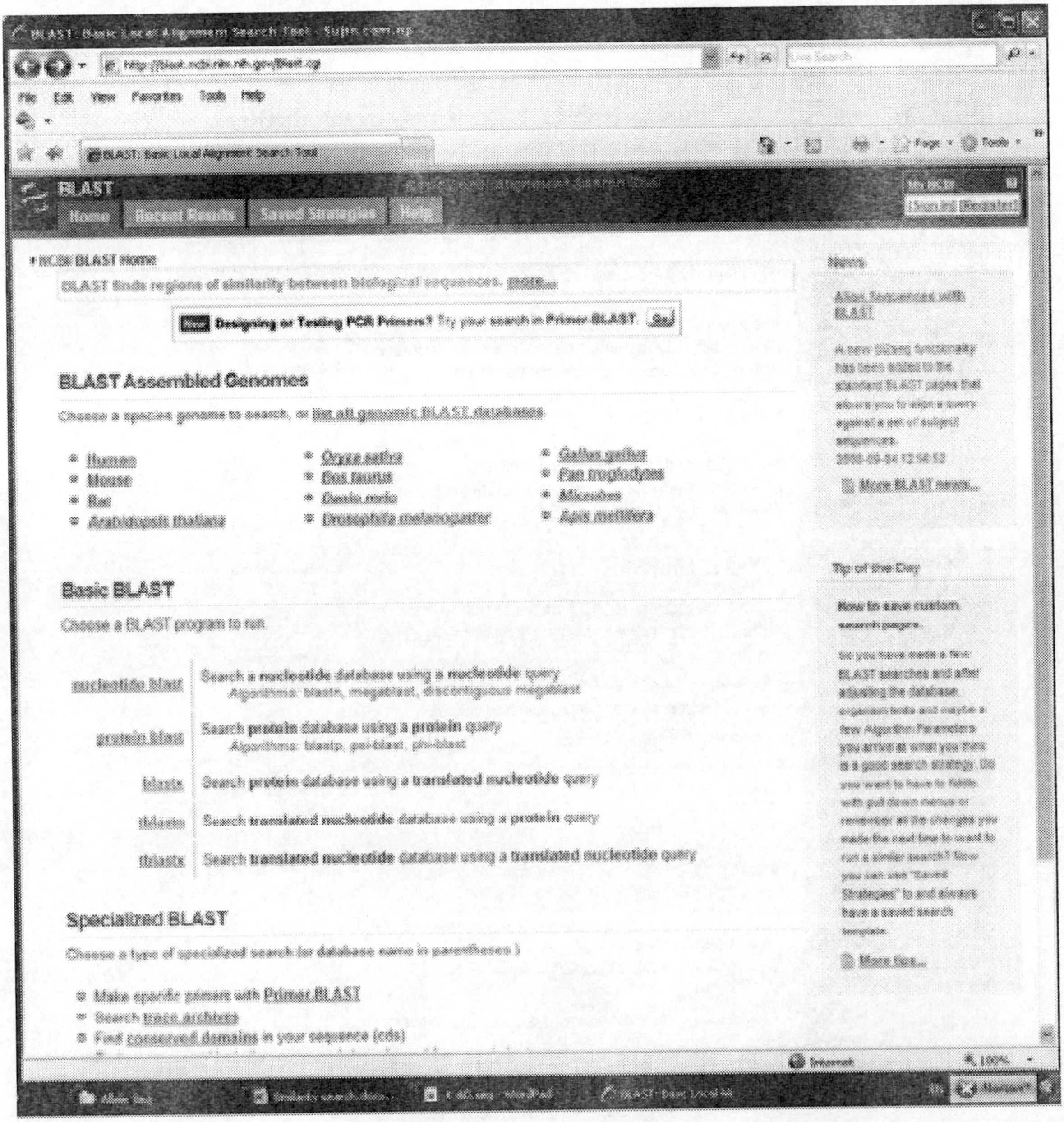

Fig. S2a. Different BLAST options.

Step 3. Paste FastA file of the query sequence in the window (Fig. S3) or browse a file from other source on the desktop from browse button. Select the database for performing BLAST from the drop down box. In this example, we have selected nr database. Then CLICK on BLAST button (left hand side of the window) to perform BLAST search.

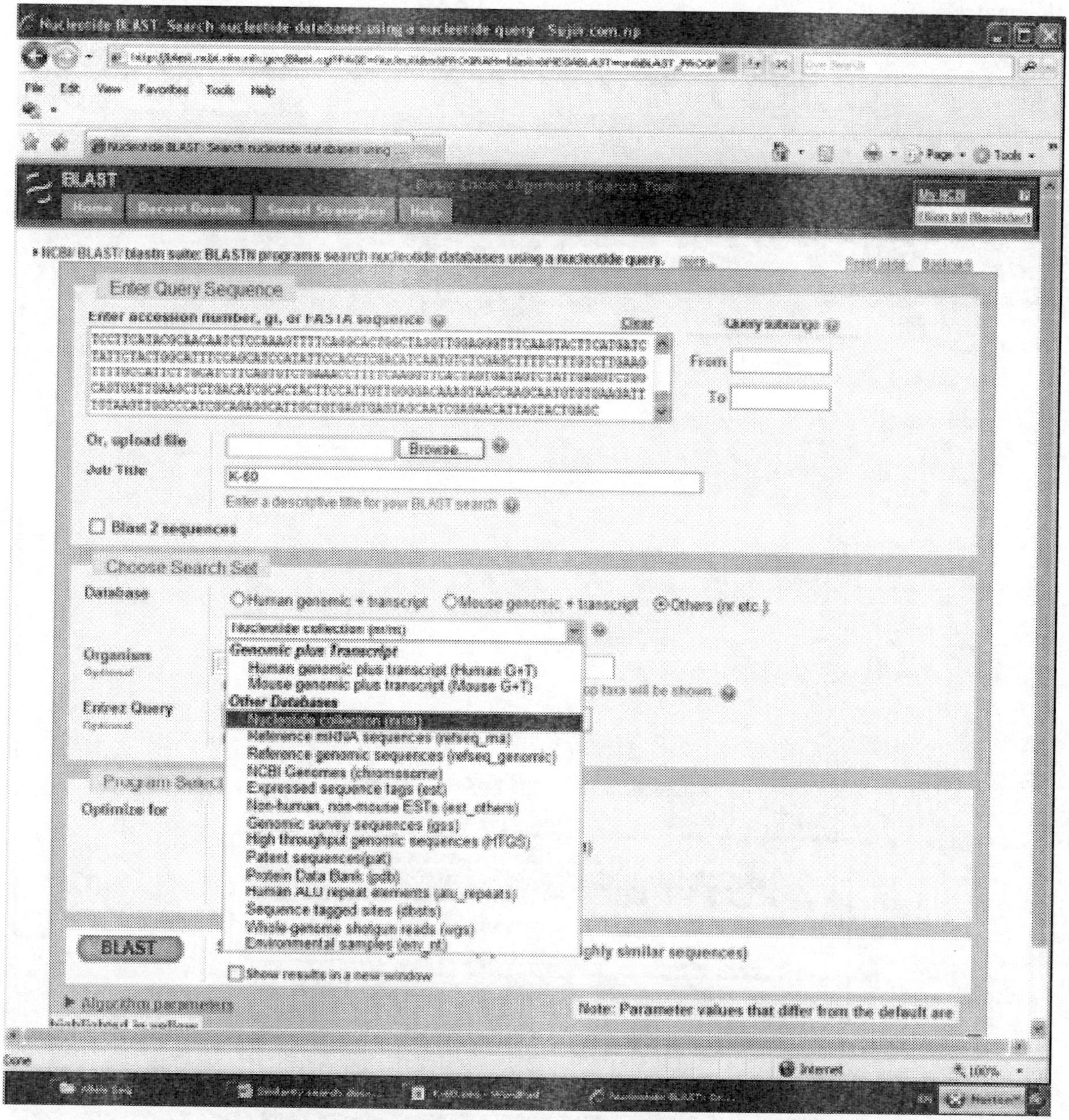

Fig. S3. BLASTn windows with query sequence pasted in it and selection of databases.

Step 4. BLAST output window will appear within a few seconds showing results of the BLAST search (Fig. S4). It will show the reference of the algorithm, size of database and links for FAQ on the top. The distribution of BLAST hits on the query sequence will show the significance of the matches.

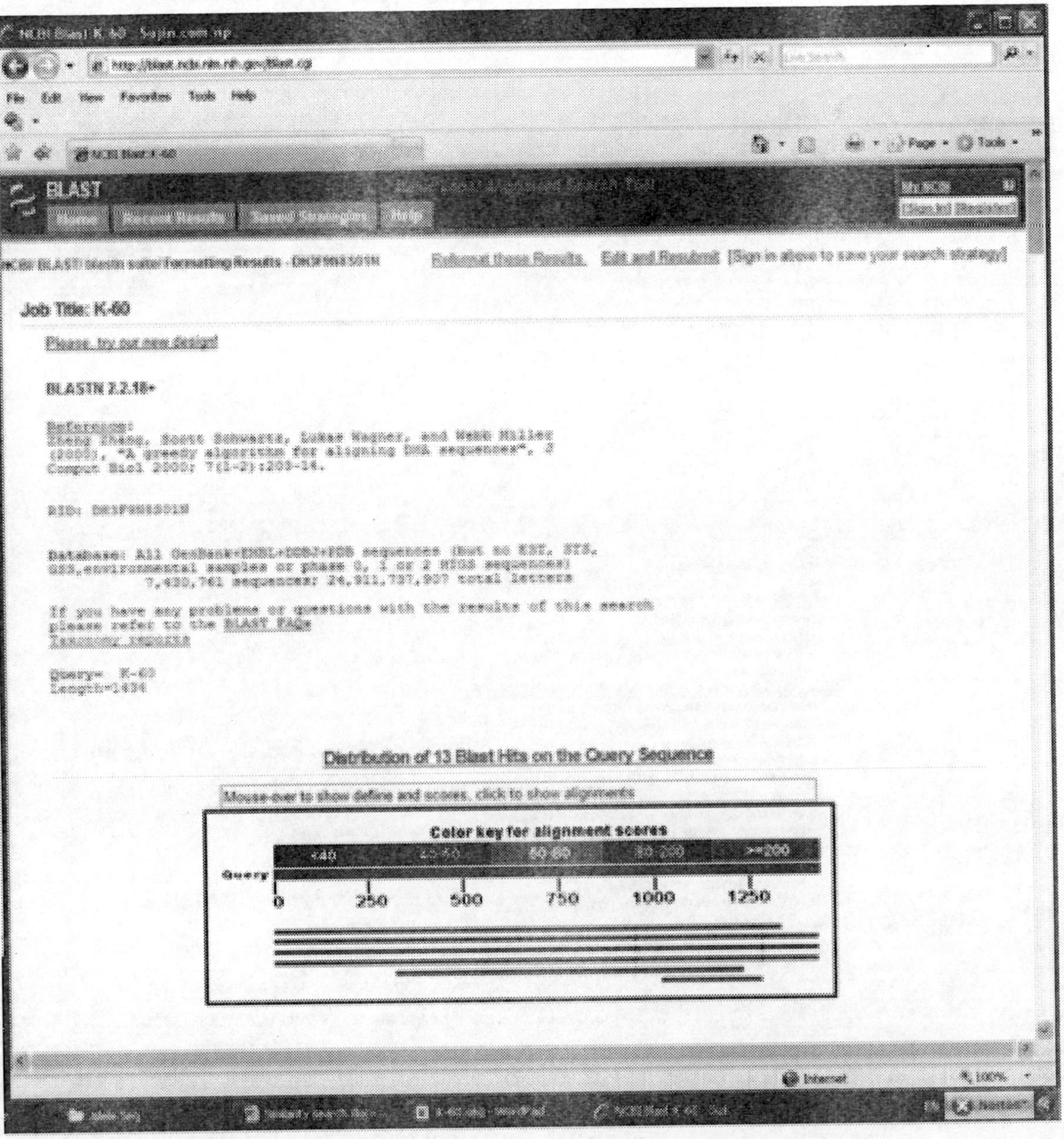

Fig. S4. Output windows showing results of the BLAST search.

Step 5. BLAST output window (bottom part). It also shows the number of hits along with their Bit score, E-values and % identities. Down below the alignment of the sequence is also shown (Fig. S5).

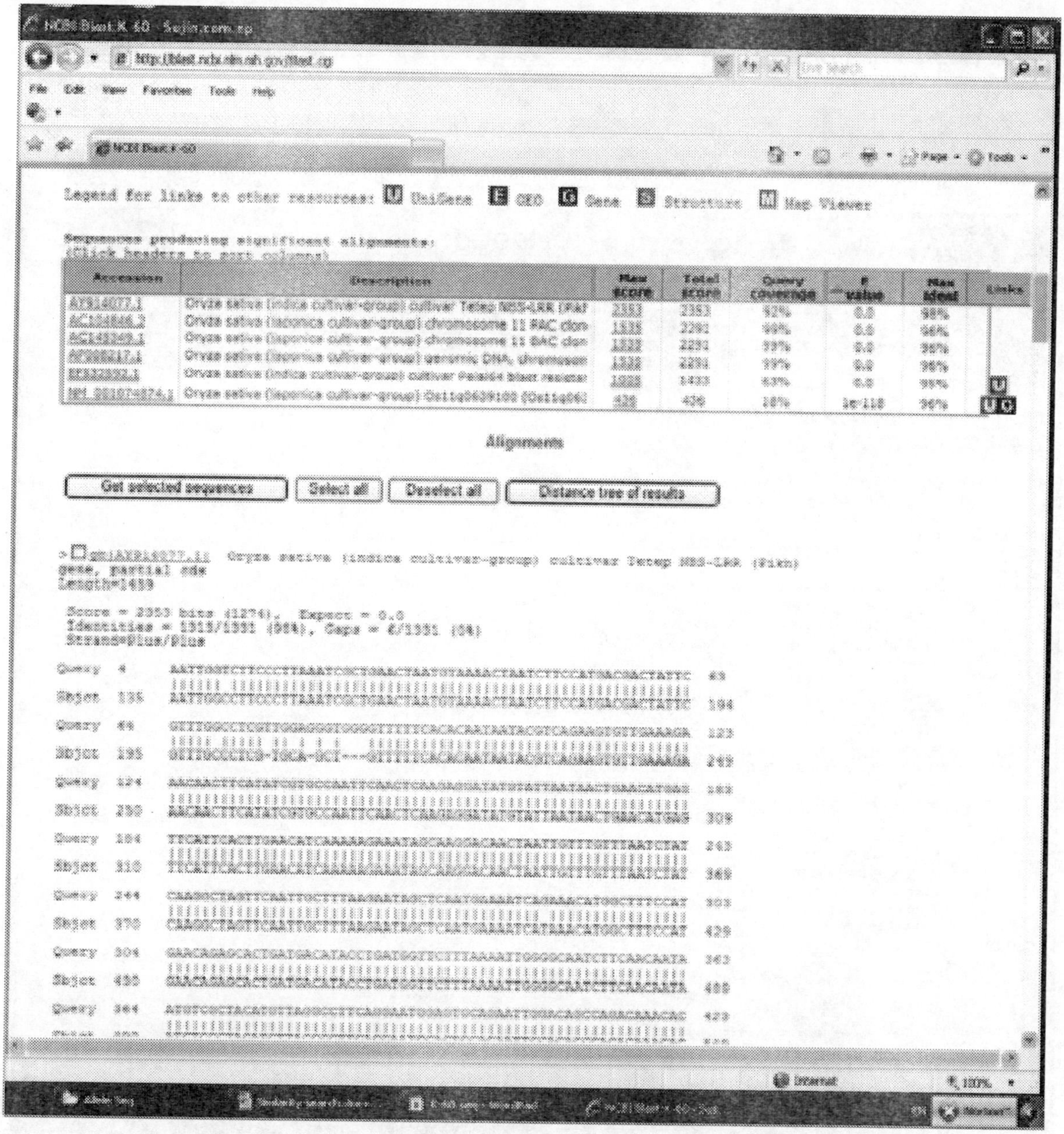

Fig. S5. BLAST output window showing the number of hits and their description.

Step 6. CLICK on the hyper link of the top hit (left hand side top) to find out the description of the match as shown in the window below (Fig. S6).

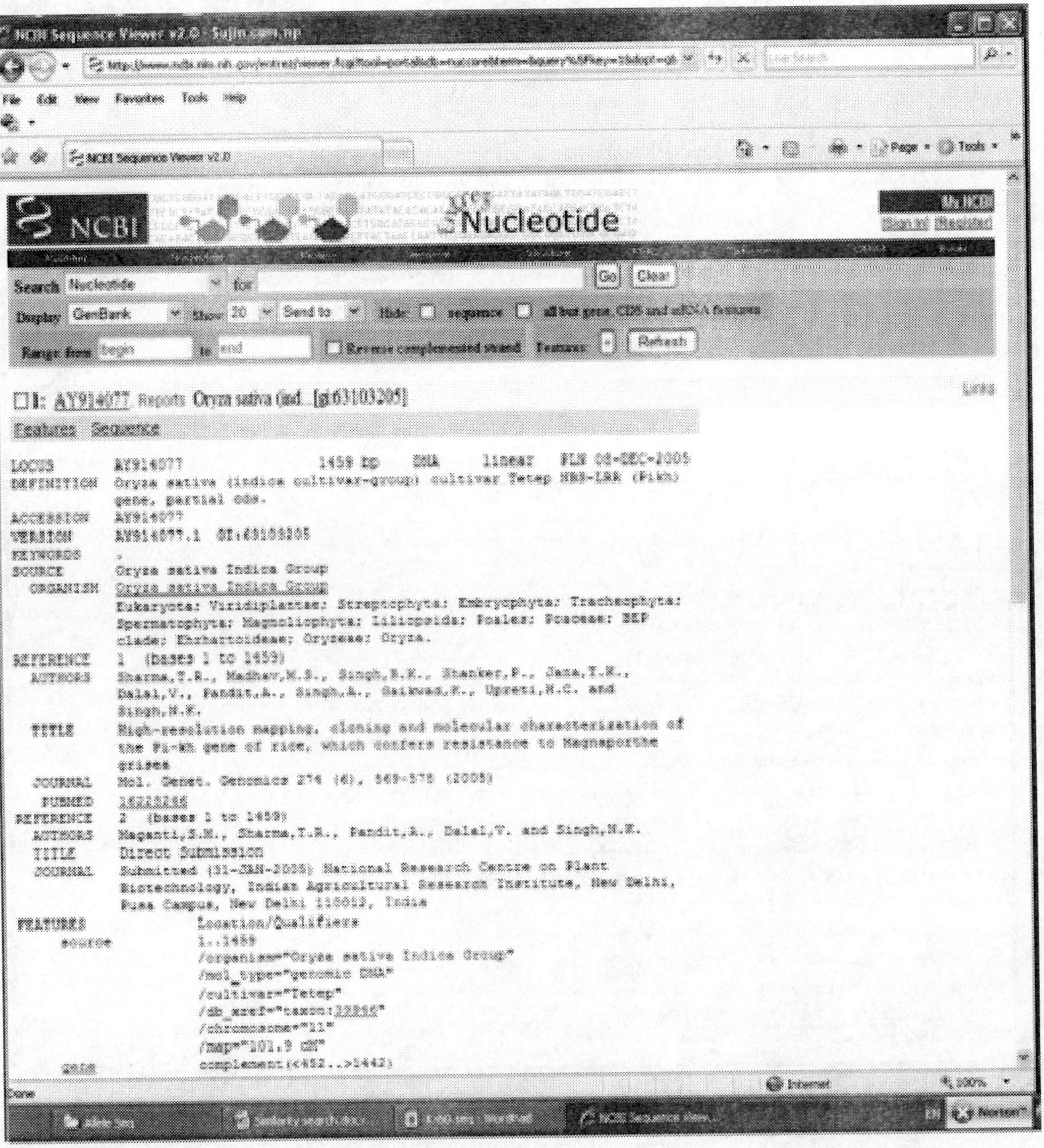

Fig. S6. Description of the matches.

Suggested Readings and Web Resources

Altschul, S.F. (1991), Amino acid substitution matrices from an information theoretic perspective. *J. of Mol. Biol.* **219**: 555-665.

Altschul, S.F. W., W Gish, W Miller, E.W. Myers and D.J. Lipman. (1990), Basic local alignment search tool. *J. Mol. Biol.* 215:403-410.

BLAST: http://blast.ncbi.nlm.nih.gov/Blast.cgi

Dayhoff, M.O., R.M Schwartz and B.C. Orcutt (1978), A model of evolutionary change in proteins." In "Atlas of Protein Sequence and Structure" 5(3) M.O. Dayhoff (ed.), 345 - 352, National Biomedical Research Foundation, Washington.

FASTA: http://www.ebi.ac.uk/Tools/fasta/index.html

Henikoff,S. and J. G. Henikoff. (1992), Amino acid substitution matrices from protein blocks." Proc. Natl. Acad. Sci. USA. 89(biochemistry): 10915 - 10919.

Henikoff,S. and J. G. Henikoff. (1993), Performance Evaluation of Amino Acid Substitution Matrices. Proteins: Structure, Function, and Genetics. **17**: 49 - 61.

Lipman, D.J. and W.R. Parson (1985), Rapid and sensitive protein similarity searches. Science **227**:1435-1441.

Pearson, W.R. (1990), Rapid and sensitive sequence comparison with FASTP and FASTA, p.63-98. In R.F. Doolite (Ed), Methods Enzymology. Vol183, Academic Press. San Diego.

7

Multiple Sequence Alignment

Sequence alignment is a very important method to find similarity among different individuals of same or different species. In the previous chapter, we have studied the sequence alignment of two DNA or amino acid sequences which is known as pairwise sequence alignment. However, if we have more than two sequences in a data set, these can be analyzed by performing (MSA) multiple sequence alignment. Therefore, MSA is defined as the alignment of at least three or more sequences.

APPLICATION OF MSA

There are many useful applications of MSA in genome analysis. However, some of the important applications are listed below:

1. It is used to find common motifs among protein families.
2. Secondary and tertiary protein structure prediction is supported by MSA.
3. MSA is used to find homology between known and unknown protein sequences.
4. It serves as the basis of phylogenetic analysis and tree construction.
5. It is used for computing consensus sequences.
6. It is used to predict new sequences falling in a given family based on patterns developed by MSA.

Factors affecting MSA

Performing multiple sequence analysis is easy by using computational tools, although the significant alignment of any given data set depends on various factors. One has to take into consideration following factors which may affect final outcome of MSA.

1. Number of sequences included in the analysis.
2. Ratio between more similar and distantly related sequences in a data set.
3. Highly divergent sequences in a data set.
4. Two sequences that are related throughout but also have divergent regions.

The error rate in the alignment is more when sequence divergence increases. It should be taken care in the beginning otherwise it goes on multiplying during subsequent steps of analysis.

Comparing multiple sequences

In multiple sequence alignment, sequences are aligned by bringing up a larger number of similar alphabets in the same column of the alignment. It can be explained as below:

Let us consider ***S*** is the set of ***k*** number of sequences of same length (without gaps) such that:

$$S = \{S_1, S_2, S_3 \ldots S_k\}$$

and ***S'*** is the set of same sequences with gaps (i.e. having different lengths) such that:

$$S' = \{S'_1, S'_2, S'_3, \ldots S'_k\}$$

The multiple sequence alignment of the sequences of set S will satisfy the condition that

$$|S'_1| = |S'_2| = |S'_3| = \ldots |S'_k|$$

$|S'_l|$ representing the length of the sequence S_l with gaps. Insertion of gaps will help in making equal length of the sequences. Thus, multiple sequence alignment can be simplified as:

$$|S_1| \neq |S_2| \neq |S_3| \neq \ldots |S_k| \xrightarrow{\text{Multiple Sequence Alignment}} |S'_1| = |S'_2| = |S'_3| = \ldots |S'_k|$$

For instance, we have three sequences i.e.S1, S2 and S3 of varying length

S1 = PLRLIR
S2 = LPIRI
S3 = PLPRLI

The multiple sequence alignment of the above sequences can be performed as below:

S1 = P L _ _ R L I R
S2 = _ L P I R _ I _
S3 = P L P _ R L I _

The information regarding most alike regions among a set of sequences can be obtained by multiple alignment. These similar regions may be referred to as conserved functional or structural domains. MSA is also helpful in the prediction of homology models of the sequences with unknown three dimensional structures. It may predict which amino acids occupy the same spatial relationship in other protein sequence of the alignment, if the structure of one or more members of the alignment is known. Multiple sequence alignment can be obtained among sequences, profiles and sequence to profile. The profile analysis is used to find motifs. Profile analysis is the process of locating sequence motifs in global MSA. Different methods used for the identification of highly conserved patterns within the larger alignments of multiple sequences help in locating motifs. The function of the genes can be predicted by aligning the unknown sequences with the sequences of known functions using MSA. MSA of the set of sequences may also be used to infer an evolutionary history of the sequences. The individuals are likely to be recently derived from the common ancestral sequences if their sequences in MSA have good alignment. Whereas, a more complex and distant evolutionary relationship is shared by poorly aligned sequences within a group.

MULTIPLE SEQUENCE ALIGNMENT METHODS

Brief descriptions of the methods used in MSA are given in the following sections. For more details readers can refer to the suggested readings given in the end of this chapter.

1. Dynamic Programming Algorithm

The Dynamic Programming Algorithms (DPA), though slow in speed, are the most accurate methods of aligning two sequences. In pair-wise alignment, two dimensional scoring matrix is created by the DPA in which each dimension represents single sequence. Each cell of the matrix has the alignment score of respective alphabets of the sequences used for the alignment. In MSA, if 3 sequences are used in the alignment, then 3-dimensional matrix will be created by DPA in order to calculate the alignment of the alphabets of all three sequences. Hence, this matrix would be a 3-dimensional cube-like structure in which each dimension will represent one sequence (Fig. 1). If large number of sequences are used in MSA, then the matrix lattice will have the number of dimensions equal to the number of sequences. In other words, in the alignment of ***n*** number of sequences, the DPA will generate an ***n*** dimensional hyper cube to calculate the optimal multiple sequence alignment. Therefore, with the increase in the number of sequences, the search space and memory required for the computation of optimal multiple sequence alignment will increase exponentially. It will result in slow speed of the alignment softwares.

2. Center-Star Method

The Center-Star method is an approximation algorithm used for the calculation of multiple sequence alignment. The optimal alignment among the multiple sequences is calculated by using

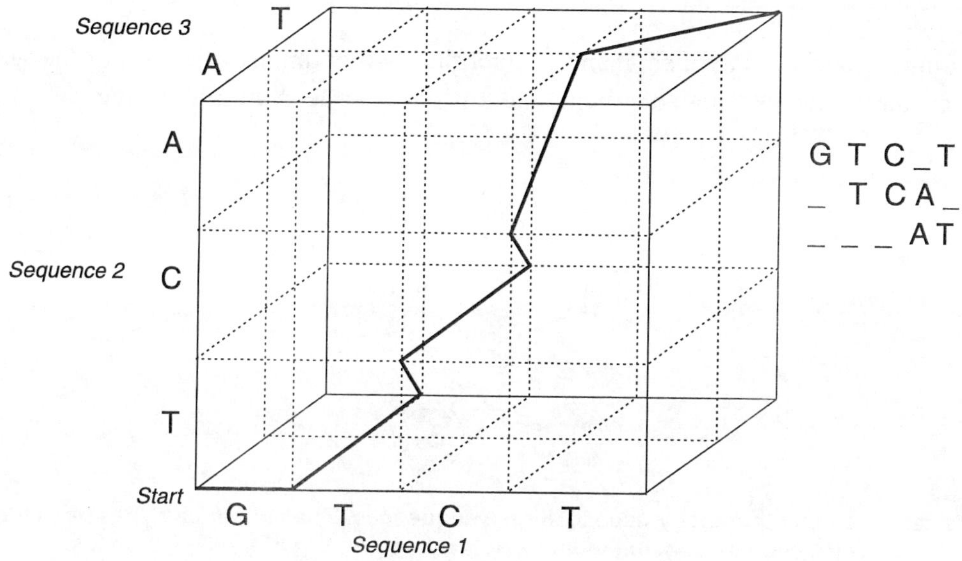

Fig. 1. Three -dimensional cube representing multiple sequence alignment algorithm.

sum of pairs rule in this method. This algorithm tries to calculate the optimal alignment with minimum sum of pairs among the sequences.

Let us again consider ***S*** is the set of ***k*** sequences such that:

$$S = \{S_1, S_2, S_3....S_k\}$$

The objective of this method is to find the optional multiple sequence alignment of the above ***k*** sequences with minimum *sum of pairs* scores.

To solve this problem, first of all define a center sequence S_c, which will be among the sequences included in a set (i.e. $S_c \in S$). This centre sequence should have minimum distances from rest of the sequences present in the data set. In other words, the center sequence can be the sequence which is closer to rest of the sequences of the experimental set. In mathematical term, ***S*** is the center sequences of Sc. Hence, S will be the sequence that minimizes distance as:

$$\Sigma_{Sj \in S} D(Sc,Sj)$$

Where S_j is any sequence from the set ***S*** except center sequences. The algorithm calculates all pairs of the sequences in the set and picks the center sequence S_c which will be close to rest of the sequences.

Once the center sequence is defined, then the algorithm will optimally align rest of the sequences with the center sequence and add them one by one to the Multiple Sequence Alignment (Fig. 2).

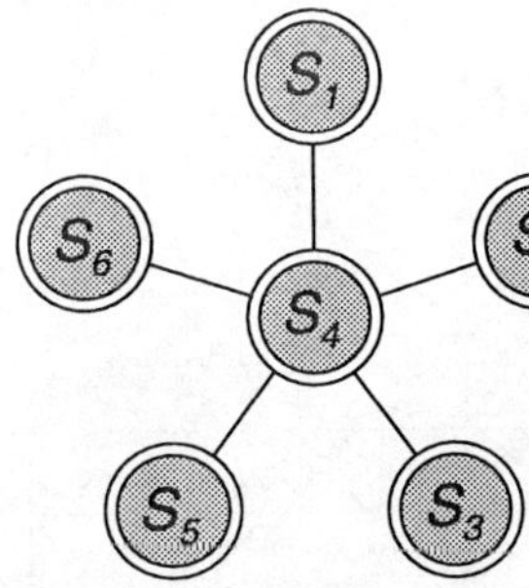

Fig. 2. The center-star methodology used to align 6 sequences. The sequence S_4 has been identified as center sequence, being most closely related to rest of the 5 sequences. Rest of the sequences are then added one by one to perform MSA.

3. Progressive Multiple Sequence Alignment

The Progressive Multiple Sequence Alignment method is based on Feng-Doolittle (1987) algorithm which is widely used to align multiple sequences. It is an heuristic approach also known as the hierarchical or tree method. It builds up the optimal MSA by combining pair-wise alignments beginning with the most similar pair and progressing to the most distantly related sequence. All progressive alignment methods have two stages.

1. First stage, in which the relationships between the sequences are represented as a tree, which is known as a *guide tree*.
2. A second stage, in which the multiple sequence alignment is built by adding the sequences one by one to the growing alignment according to the guide tree.

In this method, the MSA is achieved by computing the (*k/2*) pair-wise alignment scores (where ***k*** is the number of sequences for the alignment) and then converting these scores to distances. The incremental clustering algorithms such as UPGMA or Neighbour – Joining are used to construct tree from these distances. The most similar pairs are aligned first with the tree followed by the addition of the next most similar sequence or rest of the sequences.

The progressive multiple sequence alignments cannot be globally optimal. This is because at any stage in growing the alignment there are chances of making errors. These errors are then multiplied in the subsequent steps and reflected in the final results. If the sequences in a data set are distantly related then the performance of the alignment becomes poor. Most modern progressive methods modify their scoring function with a secondary weighting function. Based on their phylogenetic distance from their nearest neighbour, secondary weights are assigned scaling factors to individual members of the query set in a nonlinear fashion. By this way, the non-random selection of the sequences given to the alignment program is corrected. Following software are commonly used for performing progressive MSA.

A. CLUSTALW

ClustalW is the most recent version of the computational software CLUSTAL where the W means "weighting". It has the capability to provide weights to the sequences of the alignment. It is one of the important examples of Progressive Multiple Sequence Alignment. In this software, first global multiple sequence alignment is performed by computing the pair-wise alignments of all the sequences and the alignment scores. Based on alignment scores, the phylogenetic tree is computed. Then, rest of the sequences in a data set are added to the alignment one by one on the basis of the phylogenetic relationship indicated in the tree. Thus, the most closely related sequences are aligned first followed by the distantly related sequences. Similar to the FASTA or dynamic programming algorithm the initial pair-wise alignments are obtained by a fast *k-tuple* approach. The alignment scores are then converted to distances and then a guide phylogenetic tree is constructed by using different clustering methods.

ClustalW is a free ware available on http:// www.ebi.ac.uk/clustalw/. It provides many options to compute the best multiple sequence alignment by calculating sequence weights and gap penalties. Once an alignment is computed, a phylogenetic tree can be constructed to study the phylogenetic relationships among the sequences.

B. PILEUP

Another programs available in the GCG package is PILEUP. It also works on the principle of Progressive Multiple Sequence Alignment. In PILEUP, the sequences are aligned pair-wise by using Needlemann and Wunsch algorithm and the alignment scores are used to generate the guide tree by using UPGMA clustering method. The alignment of rest of the sequences is then guided by this tree. Closely related sequences are aligned first, followed by the distantly related sequences. PILEUP also does not guarantee the optimal global multiple sequence alignment.

Limitations of Progressive Multiple Sequence Alignment

The most important limitation of the Progressive MSA is the dependence of the best multiple sequence alignment on the initial pair-wise alignment of the most closely related sequences. If any error occurs during computation of the pair-wise alignment, then this error persists and is reflected in the end results. Choosing the scoring matrices and gap penalties are also the limitations of the progressive MSA.

4. Iterative Multiple Sequence Alignment

In this method pair-wise alignment scores are used as a guide to iteratively add on more additional string to a growing multiple sequence alignment. First of all, the method starts by aligning two sequences whose edit distances is the minimum in all the pairs of sequences in a data set. Then it iteratively adds the sequence which has smallest distances to any of the aligned sequences. Addition of new sequences to the multiple sequence alignment results in the insertion of gaps into all sequences of the alignment.

In many software, a variety of iterative methods of multiple sequence alignment have been implemented. For instance, the software PRRN/PRRP (http://prrn.hgc.ip) uses a Hill Climbing algorithm to optimize the alignment. Another iterative method based program is DIALIGN. It calculates local alignment between sub-segments or sequence motifs without introducing a gap penalty. The alignment of individual segments is then obtained with a matrix representation similar to DOT-PLOT matrix. Similarly, MUSCLE (http://www.arive5.com/ muscle) is another very popular package of performing MSA in which iterative methods have been used. It also uses accurate distance measures to assess the relatedness of two sequences.

5. Hidden Markov Models

Hidden Markov models (HMM) are the statistical models in which probabilistic approaches are used to assign likelihood scores to all possible combinations of gaps, matches, and mismatches to determine the most likely MSA. These models can produce a single highest-scoring output. Besides, they also generate a family of possible alignments that can then be evaluated for biological significance. Both global and local alignments can be computed by Hidden Markov Models. Significant improvements have been made in computational speed for sequences that contain overlapping regions by using HMM-based methods.

Typical HMM-based methods work by representing an MSA as a form of directed acyclic graph or partial-order graph. It consists of a series of nodes representing possible entries in the columns of a multiple sequence alignment (Fig. 3). In this representation, a column that is absolutely conserved is coded as a single node with as many outgoing connections that are possible characters in the next column of the alignment. In HMM, the individual alignment columns are the observed states and the "hidden" states are the presumed ancestral sequence from which the sequences in the query set are hypothesized to have descended. The Viterbi algorithm which is an efficient search variant of the dynamic programming method is generally used to successively align the growing multiple sequence alignment to the next sequence in query set to produce a new alignment. The HMM based alignment is distinct from progressive alignment methods since the initial alignment of the sequences is updated at the time of addition of each new sequence. However, in case of distantly related sequences, this technique can also

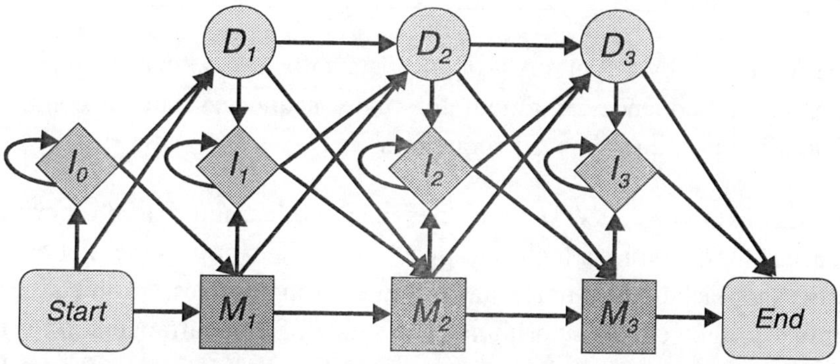

Fig. 3. The diagram is showing the model of Profile HMM to perform the global alignment of the sequences. The full model will comprise *L* layers in the model. Each layer has three states M_j, I_j and D_j. to complete the model, we add *begin* and *end* states (Silent states) to the model which are connected to layers.

***M* is the match state describing the matches in the alignment. Each *Insertion state* I_j (let I_1) has a link entering from the corresponding match state M_j (let M_1), a link towards the next match state M_{j+1} (i.e. M_2) and also has a self loop. These states can emit symbols i.e. they correspond to some character.**

To allow deletion in the alignment, the *Deletion state* $D_1, \ldots, D_j, \ldots, D_L$. These states cannot emit any symbol and called silent states. Deletion state D_j is linked with the match states M_{j-1} *and* M_{j+1}. Insertion state I_j has also link from deletion state *Dj* and towards D_{j+1}. All links carries transition probabilities of corresponding transition from one state to another state.

be influenced by the order in which the sequences in the query set are integrated into the alignment.

The variants of HMM-based methods have been implemented in several software programmes. These software are known for their scalability and efficiency in spite of the fact that using an HMM method is more complex compared to the common progressive methods. Some of the programes based on HMM are **POA** (Partial-Order Alignment) and SAM (Sequence Alignment and Modeling System).

6. Genetic Algorithms and Simulated Annealing

It is a machine learning technique used by Notredame and Higgins for the first time for performing Multiple Sequence Alignments. It is important since very high scoring multiple sequence alignments can be found with this method. This method performs MSA of the sequences in a data set by breaking the possible alignments into small fragments and repeatedly rearranging those fragments by insertion of the gaps at different positions. However, these alignments are not supposed to be optimal alignments. Hence, a general objective function (sum of pairs maximizing function) is optimized during the simulation process.

SAGA (Sequence Alignment by Genetic Algorithm), is a software tool based on genetic algorithms. It can generate multiple sequence alignment for many sequences but its computational speed slows down while aligning more than 20 sequences.

Simulated annealing (Kim *et al.*, 1994) is another machine learning based algorithm which uses the similar approaches for obtaining a high scoring MSA by the rearrangement of existing alignment using probabilistic approaches. Simulated annealing also maximizes an objective function (like sum of pairs) similar to the genetic algorithms. The speed of the rearrangement of the alignments is controlled by a temperature factor in simulated annealing method. A software tool known as MSASA (Multiple Sequence Alignment by Simulated Annealing) uses simulated annealing approach for the alignment of multiple sequences.

Some of the important software tools available for multiple sequence alignment are given in Table 1.

Table 1. List of software, their source and important features used for MSA.

S. No.	Software	Source	Features
1.	MAFFT v 6	http://align.bmr.kyushuu.ac.jp/mafft/online/server/	It can align large number (>20000) of unaligned sequences and can perform rough clustering using N-J and UPGMA approach. It can perform MSA by progressive as well as iterative approaches.
2.	MUSCLE	http://www.drive5.com/muscle/ http://www.ebi.ac.uk/Tools/muscle/index.html	MUSCLE is a free ware used for protein and nucleotide sequences alignment. MUSCLE stands for multiple sequence comparison by log-expectation. It works on the basis of iterative alignment methods and improves on progressive methods with a more accurate measurement to assess the relationship between two sequences.
3.	CLUSTAL W	http://www.ebi.ac.uk/Tools/clustalw2/index.html	ClustalW2 is a general purpose multiple sequence alignment program for DNA or proteins sequences. It produces biologically meaningful multiple sequence alignments of divergent sequences. It calculates the best match for the selected sequences, and lines them up so that the identities, similarities

Contd....

S. No.	Software	Source	Features
			and differences can be seen. Evolutionary relationships can be seen via viewing Cladograms or Phylograms in this software.
4.	MAP	http://genome.cs.mtu.edu/map.html	It can align both DNA as well as amino acid sequences. It provides the options for input sequences or files in FASTA format and setting the alignment parameters like match, mismatch scores, gap costs etc.
5.	DIALIGN	http://dialign.gobics.de/ http://bibiserv.techfak.unibielefeld.de/dialign/	It is an iterative alignment based tool freely available to download. It can align both DNA and protein sequences. Sequences can be given as input either in FASTA format or a FASTA file. DIALIGN constructs pair-wise and multiple alignments by comparing entire *segments* of the sequences. No gap penalty is used in this software. This approach can be used for both global and local alignment, but it is particularly successful in situations where sequences share only *local* homologies.
6.	DCA	http://bibiserv.techfak.unibielefeld.de/dca/	Divide-and-Conquer Multiple Sequen-*ce Alignment* (**DCA**) is a program for producing fast, high quality simultaneous multiple sequence alignments of amino acid, RNA, or DNA sequences. The program is based on the DCA algorithm, which is a heuristic approach to sum-of-pairs (SP) optimal alignment.
7.	HMMER	http://hmmer.janelia.org/	HMMER is a freely distributable implementation of profile HMM software for protein sequence analysis. It is based on profile hidden Markov models which can be used to do sensitive database searching using statistical description of a sequence family's consensus.
8.	MEME	http://meme.sdsc.edu/meme/website/intro.html	It is based on Expectation Maximiza-tion method. It searches for motifs and then queries them against the database.

EXERCISE

Multiple Sequence Alignment using ClustalX

Clustal is an important software to perform multiple sequence alignment of DNA or Protein sequences. In this example an alignment of few amino acid sequences has been performed by using ClustalX program. Following steps should be followed for using this program:

Step 1. Open ClustalX program by double clicking on ClustalX icon on the desktop window.

Step 2. After opening the ClustalX window, load the input fasta file of the sequence by using **Load Sequences** option in the **File** menu. (Fig. S1)

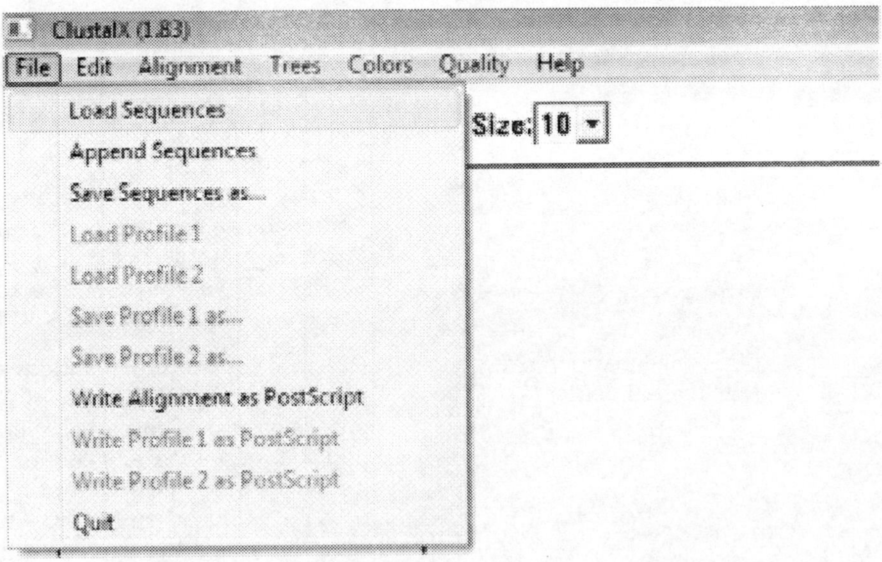

Fig. S1. Loading Sequences from the word pad file.

Step 3. As we click on **Load Sequences,** a file browsing dialog box open from there input file is selected. Select the sequence file and click on open to load the sequence in ClustalX (Fig. S2).

Step 4. As we click on OPEN, the sequence file is loaded in ClustalX programme (Fig. S 3). To perform alignment, click on **Do Complete Alignment** under **Alignment** menu (Fig. S3).

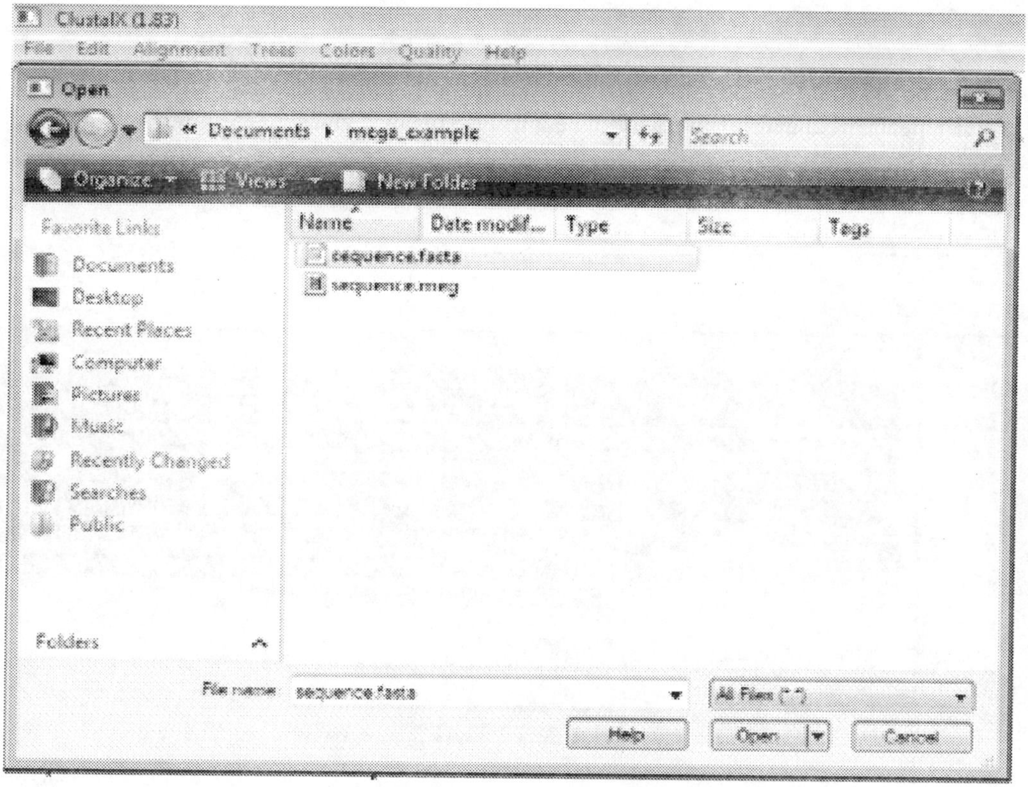

Fig. S2. Browsing Sequence File stored in fasta format.

Step 5. When we click on **Do Complete Alignment** option, it opens a Complete Alignment dialog box to select the Output Guide tree file and output Alignment file (Fig. S4).

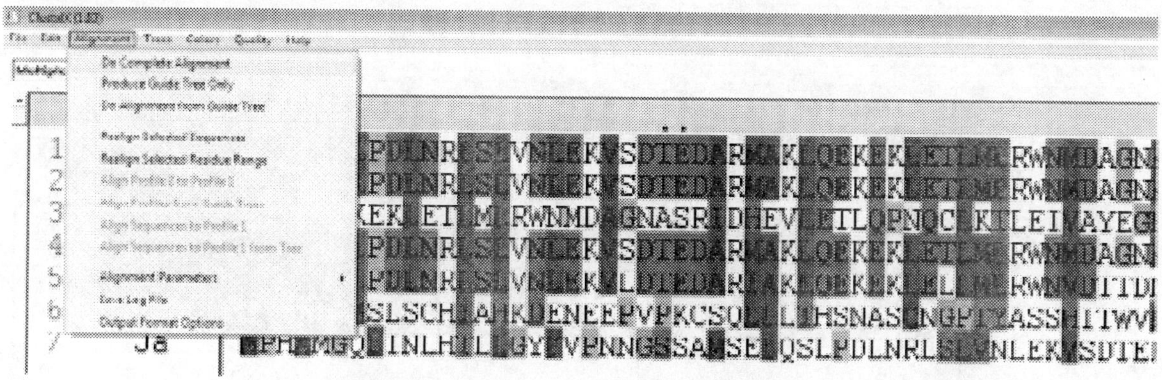

Fig. S3. Input file loaded in ClustalX and selection of Alignment menu.

Step 6. Click on **ALIGN** button in **Complete Alignment** dialog box as shown in Figs. S4. The ClustalX will perform the multiple sequence alignment and gives the aligned sequences as an output (Fig. S5).

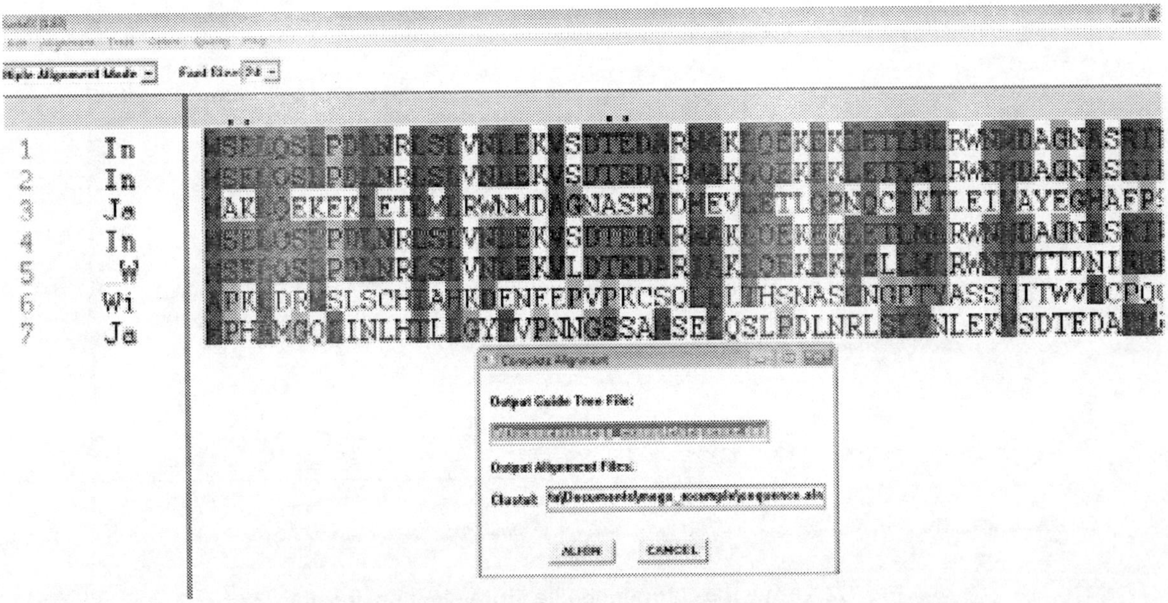

Fig. S4. Complete Alignment dialog box and guide tree output options.

Fig. S5. Sequences after complete Multiple Sequence Alignment. *shows the completely aligned sequences, whereas – are inserted as gaps.

Suggested Readings

Baxevanis A.D. and B.F.F Ouellette (2001), Bioinformatics: A practical guide to the analysis of genes and proteins. John Wiley & Sons, Inc. NY, USA.

Brown S.M. (2000), Bioinformatics: A biologists guide to computing and the internet. A Biotechniques Books Publication, Eaton Publishing.

D.-F. Feng and R.F. Doolittle (1987), Progressive sequence alignment as a pre–requisite to correct phylogenetic tree. J Mol Evol. **25**: 351-360.

Haubold, B. and T. Wiehe (2006), Introduction to computational biology: An evolutionary approach. Birkhauser Verlag, Basel-Boston-Berlin, Germany.

Thompson J.D., D.G. Higgins and T.J. Gibson (1994), CLUSTAL W: improving the sensitivity of progressive multiple sequence alignment through sequence weighting, position-specific gap penalties and weight matrix choice. *Nucleic Acids Res.*, **22**: 4673-4680.

8

Phylogenetic Analysis

The study of evolutionary relatedness among various groups of organisms (e.g., species, populations) is known as phylogenetics. Phylogenetics derives from the Greek words **phyle** = tribe, or race and **genetikos** = relative to birth (from genesis = birth). It is also known as systematics or cladistics in which a species is treated as a group of individuals with common lineage over time. The basic idea of phylogenetic analysis is to compare species based on specific *characters* (features) with the assumption that species with similar characters are genetically related to each other. These relationships are referred to as phylogeny and are represented in the form of a phylogenetic tree.

The morphological markers e.g. in plants like leaf size, color, number of branches, etc are generally used for phylogenetic analysis. However, DNA and protein sequences are used for phylogeny studies in molecular biology. Usually, long strings of DNA or protein sequences are used as a character for analysis. The conserved alignment of several sequences is used to deduce relationship among species.

A phylogenetic analysis of a family of related DNA or protein sequences is the process of determining the derivation of family during the evolutionary course. The evolutionary relations among the sequences can be predicted by placing the sequences on the outer branches of a tree. The inner branches of the tree reflect the degree to which different species are related to each other. Two sequences which are very much similar to each other will be located as neighboring outside branches and will be joined to a common branch beneath them.

Phylogenetic analysis of nucleic acid and protein sequences has been a very important area of sequence analysis. On the basis of phylogeny, most closely related sequences occupy neighboring branches on a phylogenetic tree. Phylogenetic relationships among the genes can be helpful in the prediction of their functions which are similar in a gene family of an organism. It can also be used to follow the mutation occurring in the rapidly changing species, such as virus.

Multiple sequence alignments and phylogenetic analysis methods are strongly linked. Using these methods, two very much similar sequences can be easily aligned by looking at them. Hence, group of much similar sequences having a small level of variation throughout the sequence can be easily depicted in a tree. However, it is much more difficult to align sequences which are evolutionary different because of so many possible evolutionary paths that could be followed to produce the observed sequence variation, it is very difficult to perform phylogenetic analysis. The conservation of sequences at different positions can be observed from the sequence alignment. The two sequences are called homologous, if they share a common evolutionary relationship.

WHAT IS A PHYLOGENETIC TREE?

It is assumed that all organisms had a common ancestor. During the course of evolution, a species splits into two or more different species that do not cross among each other. The relationship among different species can be represented in the form of an evolutionary or phylogenetic tree (Fig 1). In this figure **L, M, N, O, P** are the leaf nodes or external nodes or termini of the branches. These are also called **Operational Taxonomic Units (OTUs)**. These can be gene or protein sequences, individuals, populations, species etc. **Q, R** and **S** are internal nodes which represent ancestral units. **T** is the root node i.e. most ancestral node. The order of the nodes on the tree is known as tree topology.

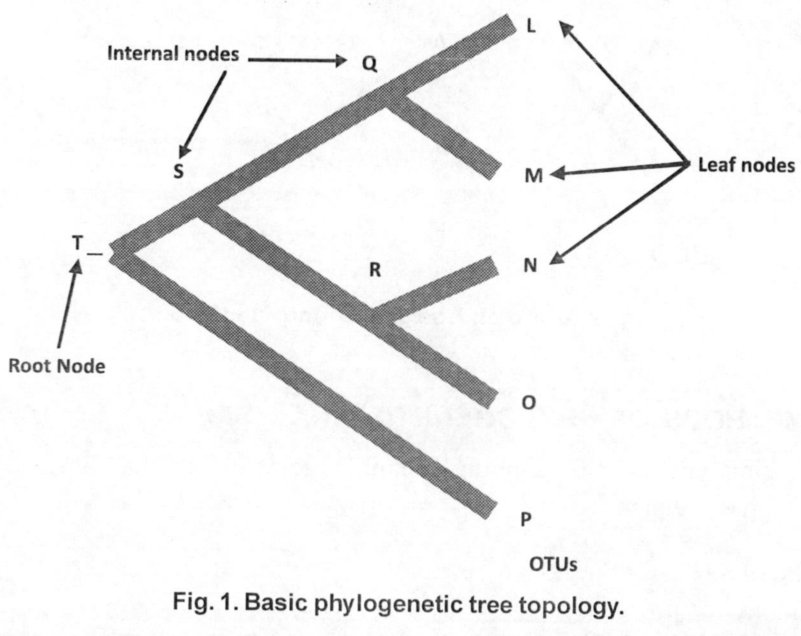

Fig. 1. Basic phylogenetic tree topology.

Phylogenetic tree is a two-dimensional graph which takes several forms. They can be ***rooted*** or ***unrooted***, binary or general, and may show, or may not show edge lengths (Fig.2.). A *rooted* tree is a tree in which one of the nodes is connected to the root, which determines the direction of the ancestral relationship. An *unrooted* tree has no pre-determined root and it only shows how close (or distant) are the species. Therefore, in case of unrooted tree, the distance between the nodes should be symmetric (since the tree edges are not directed). An unrooted tree can be converted into a rooted tree by inserting a new node, which will function as the root node. Hence, an out-group is defined in the data set for the construction of a tree. The out-group is a species that is definitely different from all the species of interest. The proposed root will be the direct predecessor of the out group. A rooted and unrooted tree is shown in Fig. 2. A **binary** or ***bifurcating*** tree is a tree in which a node may have only 0 to 2 branches, that is, in an unrooted tree, there would be up to three neighbours.

The branch in a tree shows the genetic distance between the connected nodes. The existence of a *molecular clock* is also assumed. The molecular clock is a constant pace of the evolutionary process. It can be explained theoretically by producing a phylogenetic distance-preserving tree which can be presented along a time-axis. Each node is assigned the time in which it happened in the course of evolution. In such a perfect tree, the difference in time between the parent node and the child node form the length of each edge.

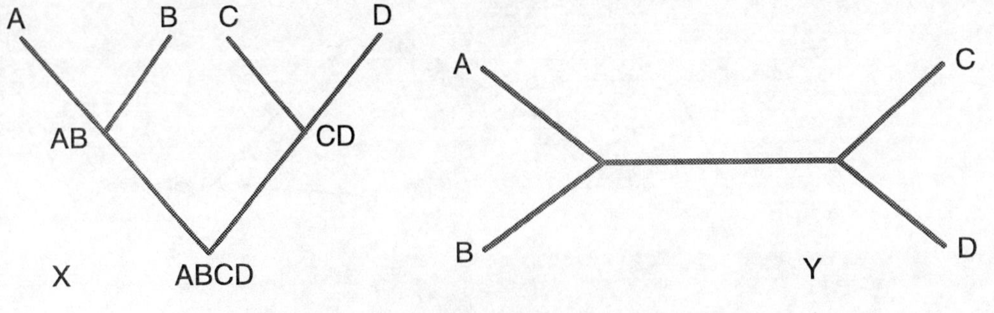

Fig. 2. X Rooted Tree and Y– Unrooted Tree.

DIFFERENT METHODS OF PHYLOGENETIC ANALYSIS

There are three main methods for finding the phylogenetic tree that best accounts for the variation in a group of sequences.

Distance based Methods
Maximum Parsimony
Maximum Likelihood

1. Distance based methods

Many methods are used to solve the phylogeny problems based on distances. The phylogenetic tree which is produced based on the number of changes between each pair in a group of sequences is one of the most important methods. The basic objective of distance methods is to find a tree in which the nearest neighbour are positioned correctly. The branch length of tree should represent the original data as closely as possible. A successful distance based method is based on the degree to which the distances among a set of DNA or protein sequences can be made additive on a predictive evolutionary tree. The sequence pairs that have smallest numbers of mismatches are known as neighbours. The major advantage of this method is that it is the simplest method of tree building which is based on most simple clustering method.

ALGORITHMS FOR CLUSTERING USING DISTANCE BASED METHODS

A. Unweighted Pair Group Method with Arithmetic Mean

Unweighted Pair group Method with Arithmetic Mean (UPGMA) is a clustering algorithm that works by joining the branches of a tree on the basis of greatest similarity criteria among pairs of sequences and by calculating means of joined pairs. This method has become very popular because of the ease and simplicity with which it is used. It first calculates the raw pair-wise distance data from a set of sequences, then constructing a matrix. The algorithm first uses the least distance pairs for grouping and then adds sequences having distance next to it and so on to build a tree from all the sequences in a data set. This method works on the assumptions that; i) the rate of change along the branches of the tree is constant, ii) it calculates branch length between most closely related sequences then takes mean of the distance between these pairs, iii) the analysis continues until all the sequences are included in a tree and iv) position of the root of the tree is predicted in the final step. The final output of the algorithms is the construction of a tree (Fig.3).

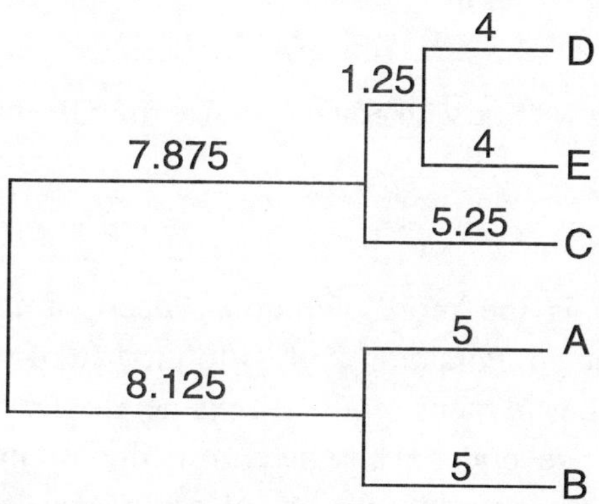

Fig. 3. A simple example of the tree constructed by using UPGMA method. The name of species (A-D) are given on the termini of branches. The numbers aove the branches are showing branch length.

B. Neighbor-Joining Method

The Neighbor-Joining (NJ) method is very fast and greedy heuristic in which the closest sub-tree are first joined to each other followed by the joining of sub-trees far from each other. This method is based on minimum evolution principle. It is most suitable for the trees of known topology having branch length which simulates different levels of evolutionary changes. In each iteration, the algorithm tries to find out the direct ancestor of two species in the tree. The tree construction starts i) without giving any preference to the pairing of sequences, ii) it combines pairs of sequences by finding a pair that will minimize the branch length, and iii) it creates new distance table by using Fitch-Margoliash algorithm. The trees generated by this method are shown in Fig. 4 A&B.

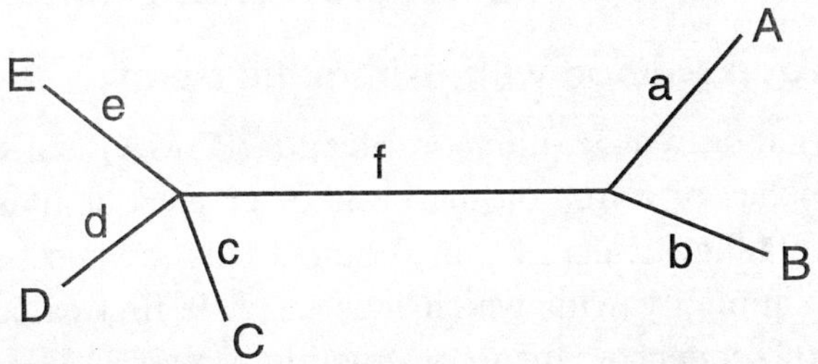

Fig. 4A. NJ tree with unequal distance from the nodes.

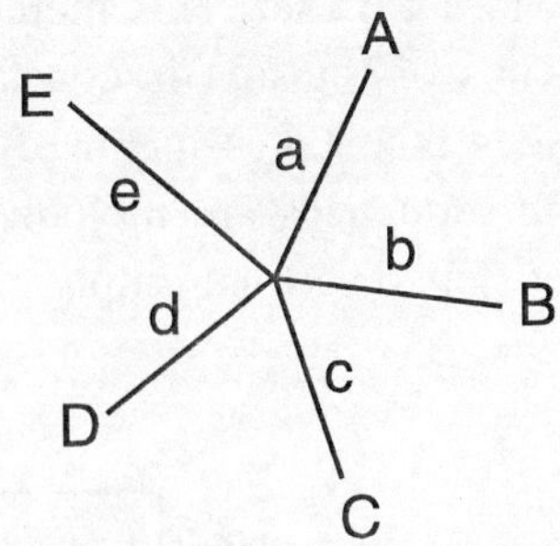

Fig. 4B. A NJ tree with equal distance from the nodes.

C. Maximum Parsimony

Maximum parsimony method is the most commonly used in those cases where ancestral relationship has to be reconstructed. Parsimony refers to the use of simple answer to a particular problem. It minimizes the number of steps required to generate the observed sequence variation. For any particular site, there are several ways to determine the minimum number of evolutionary events. For the execution of this method on a set of sequences, multiple sequence alignment (MSA) is required in which positions of the sequences corresponding to each other are predicted. A phylogenetic tree is constructed for each position in MSA based on the smallest number of evolutionary changes. Each possible tree is evaluated by using this method to give a specific score so that best tree can be selected from the different trees. The tree with minimum number of evolutionary changes is also called the most parsimonious tree. If the number of substitutions per site is small, parsimony does not require exact constancy of rates of change between the branches. However, if the total length of examined sequence is small and there are a large number of backward and parallel mutations then erroneous tree may be produced by the parsimony method. This method is very good for distantly related sequences.

D. Maximum Likelihood

Maximum Likelihood method is based on the explicit model of evolution used for phylogeny analysis. It searches the phylogenetic tree and evolutionary model based on the highest likelihood of producing the observed data set. The method assumes that a history with higher probability is preferred over the history of lower probability to get the observed state to construct a tree with the highest probability or likelihood. This method starts with an evolutionary model which provides the estimates of the rates of substitution of one base to another in DNA/protein sequences. All possible trees of the given number of sequences are listed by this method. The sequence data for the first site i.e. first column of multiple sequence analysis is put on the outer leaves of the trees. The probability of each tree is calculated which is the product of mutation rates of each branch of the tree. This analysis is repeated for all the sites in the multiple sequence alignment and a most likely tree with the highest probability is generated. The maximum likelihood methods has many advantages over other methods, (i) these have lower variance compared to other methods, (ii) very robust, (iii) better for very short sequences, (iv) statistically well founded, (v) evaluate different tree topology and (vi) uses all the sequence information. However, the biggest disadvantage of this method is that it is extremely slow and need more computational power and the results are dependent on model of evolution.

PHYLOGENETIC ANALYSIS SOFTWARE

Phlogenetic trees can be constructed with the help of various computational tools. Many software are now available to perform online analysis of the sequence data on world wide web. Some of the important software, their web address and description are given in Table 1.

EXERCISE

Phylogenetic analysis of amino acid sequences using Mega software

The Mega 4 software was first downloaded from http://www.megasoftware.net and installed on a desktop. An example of the amino acid sequences of an allele amplified from different rice species

Table 1. Software available for Phylogenetic Analysis on World Wide Web.

S. No.	Name of the Software	Website address	Description of the software
1.	**Winboot**	http://www.irri.org/science/software/winboot.asp	In Winboot software package, the UPGMA trees are calculated from binary (0/1) data. Each group of the trees is supported by bootstrap values by using bootstrap analysis. The program can read binary data in 0/1 form prepared in excel sheet or in PHYLIP format. Simple similarity coefficients computed by this program are also used to carry out bootstrapping for each input file. The final output of the analysis is obtained in the form of a consensus tree.
2.	**Phylip**	http://evolution.gs.washington.edu/phylip.	PHYLIP is available free, from its web site, in C source code or as executables for Windows, Mac OS 8 or 9, and Mac OS X. The C source code can be easily installed on UNIX or Linux systems. It includes programes to carry out parsimony, distance matrix methods, maximum likelihood, and other methods on a variety of types of data, including DNA and RNA sequences, protein sequences, restriction sites, 0/1 discrete characters data, gene frequencies, continuous characters and distance matrices.
3.	**Mega**	http://www.megasoftware.net	MEGA (Molecular Evolutionary Genetic Analysis) software molecular data is analyzed by using parsimony, distance matrix and likelihood methods resulting in a consensus tree along with bootstrapping. A variety of distance measures, with Neighbor-Joining, Minimum Evolution, UPGMA, and parsimony tree methods are used in this software. A variety of data editing tasks such as tests of the molecular clock, and single-branch tests of significance of groups are also performed by it. The latest version of the software is MEGA 4.
4.	**TreeCon**	http://bioinformatics.psb.ugent.be/psb/Userman/treeconw.	TREECON is a software package for the construction phylogenetic tree by using distance data. Several equations are included to convert dissimilarity into evolutionary distance and several methods (such as neighbor-joining) are included for inferring the tree topology. It also includes bootstrap analysis. It also has good facilities for drawing trees. The program is freely available and runs on PCs under Windows.

has been used for analysis. Phylogenetic analysis was performed by using Mega 4 software from the FASTA format of sequences. Steps used in performing this analysis are given below:

Step 1. Open MEGA software by clicking on the icon of Mega on the desktop.

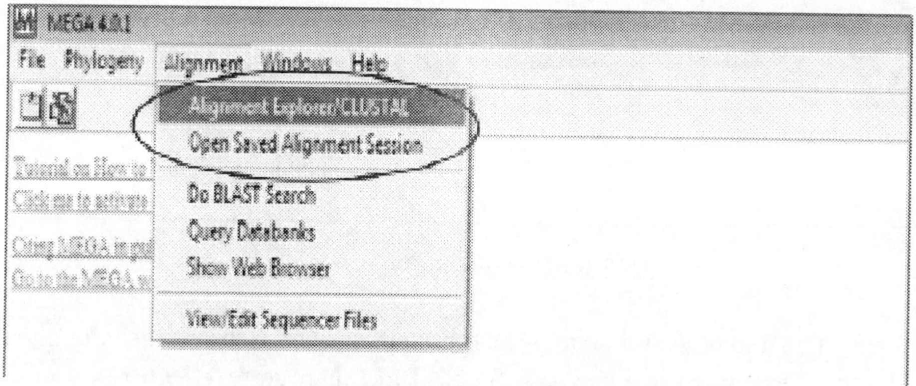

Fig. S1. Mega Window and Alignment Explorer Options.

Step 2. The Mega 4.0.1 window will appear on the screen as in figure S1. Select Alignment Explorer/CLUSTAL from the alignment menu.

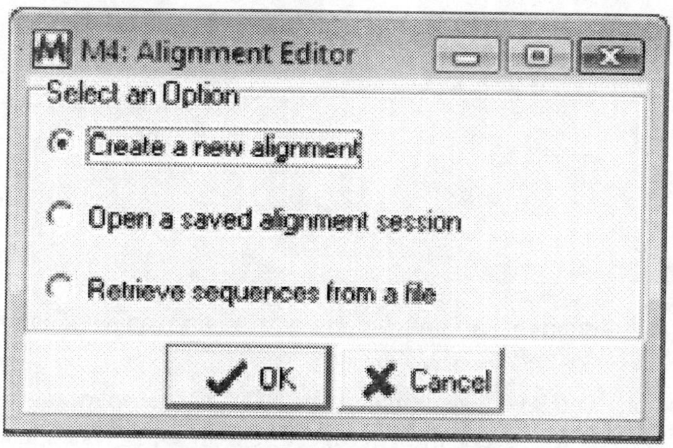

Fig. S2. Alignment Editor with select options.

Step 3. An alignment dialog box will open to select an option. Select "Create a new alignment" from the list and click on **OK**. (Fig. S2)

Step 4. As we select "Create a new alignment" option, a new dialog box will open asking for type of input sequence. Click on **YES**, if the input sequences are nucleotides and **NO** for protein sequences. We clicked on **NO** as we are using amino acid sequences in our example. (Fig. S3).

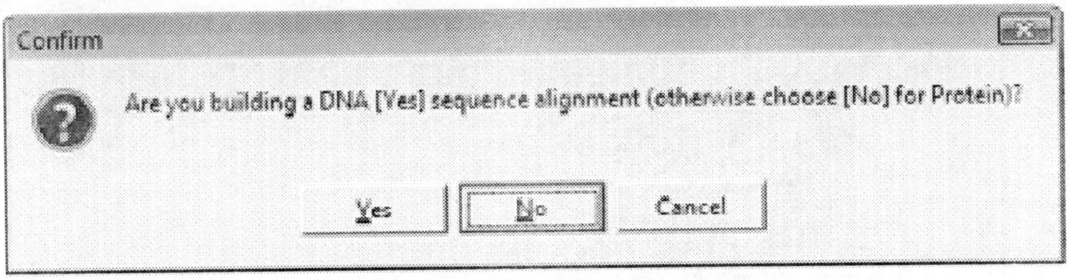

Fig. S3. Input sequence type selection.

Step 5. As we click on NO option to select protein sequence type, the alignment explorer window will appear as shown in figure S4. Select **Data** then go to >> **OPEN and** click on >> **Retrieve sequences from file** to load the input sequences.

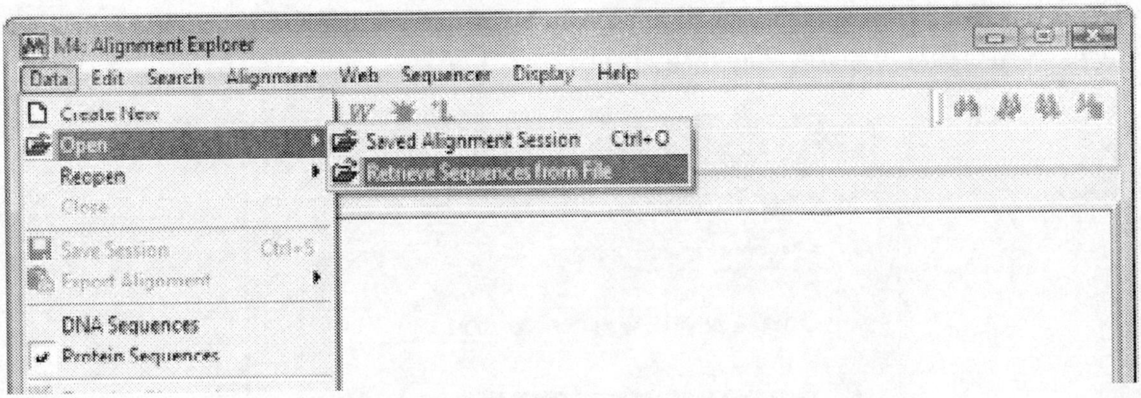

Fig. S4. Alignment Explorer window and loading input sequences.

Step 6. As we click on **Retrieve Sequence from File,** a file browsing dialog box will open to select the input FASTA file. Select the file and click on OPEN as shown in Fig. S5.

Step 7. Sequences will open in Alignment Explorer as given in Fig. S6. Select "Align by ClustalW" from the Alignment menu. A Dialog box will appear from which select all sequences.

Step 8. The "ClustalW Parameters" dialog box will open to select the alignment parameters. Click on OK after selection of the parameters (Fig. S7).

Step 9. After alignment, save the alignment session for further uses and save the alignment in MEGA format for further analysis using MEGA software. (Fig. S7, S8 and S9). MEGA software asks for the title of the data for saving in MEGA Format.

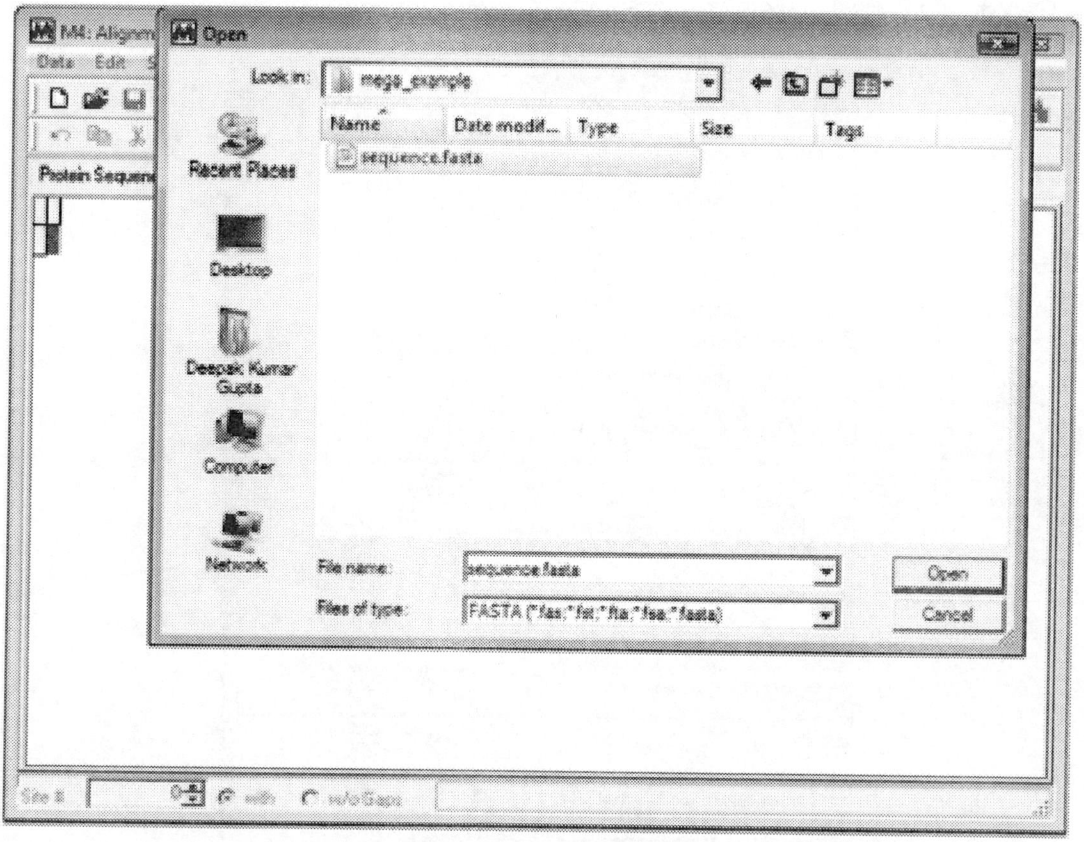

Fig. S5. File browsing window.

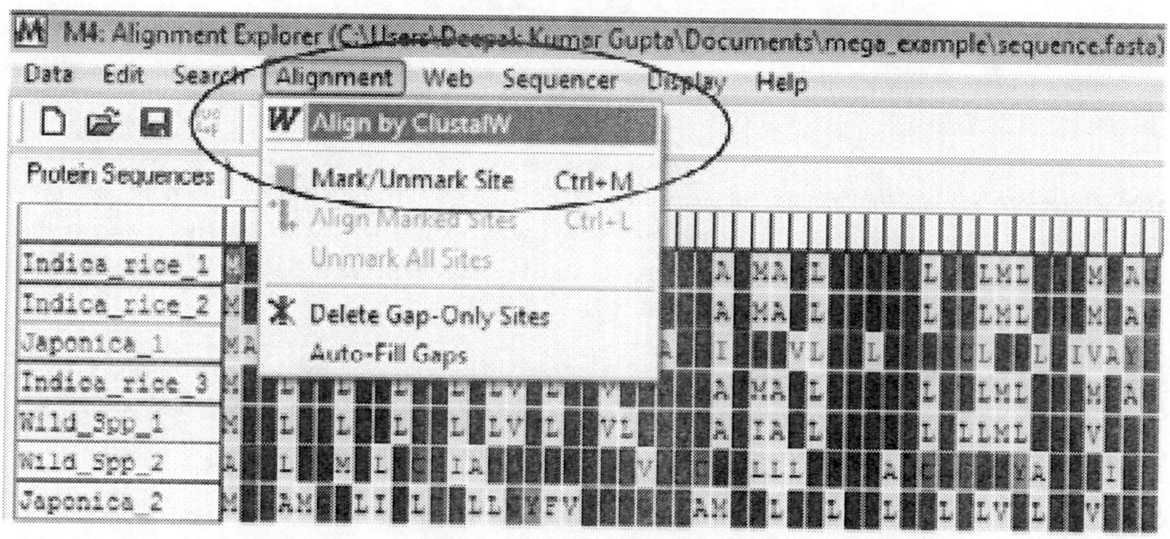

Fig. S6. Alignment Explorer.

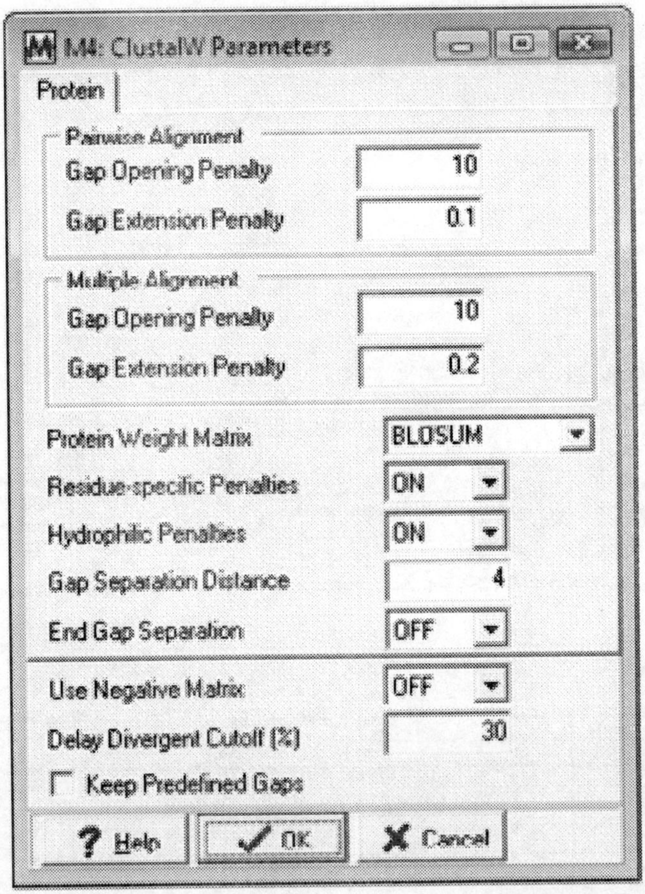

Fig. S7. ClustalW Parameters Dialog Box.

Step 10. After saving the alignment session and alignment file, close the alignment explorer window. As we close the alignment explorer window, a dialog box will open to confirm the opening of the data file in MEGA software for further analysis (Fig. S10). Click on **YES** to open the data. Data will be opened in MEGA as in Fig. S11.

Step 11. To perform phylogenetic analysis of the data, select any of the options from PHYLOGENY menu as per requirement. Here, we selected **PHYLOGENY >> Bootstrap test of Phylogeny >> UPGMA** for phylogenetic tree construction (Fig. S12).

Step 12. Select the parameters from analysis preferences window and click on COMPUTE (Fig.S13).

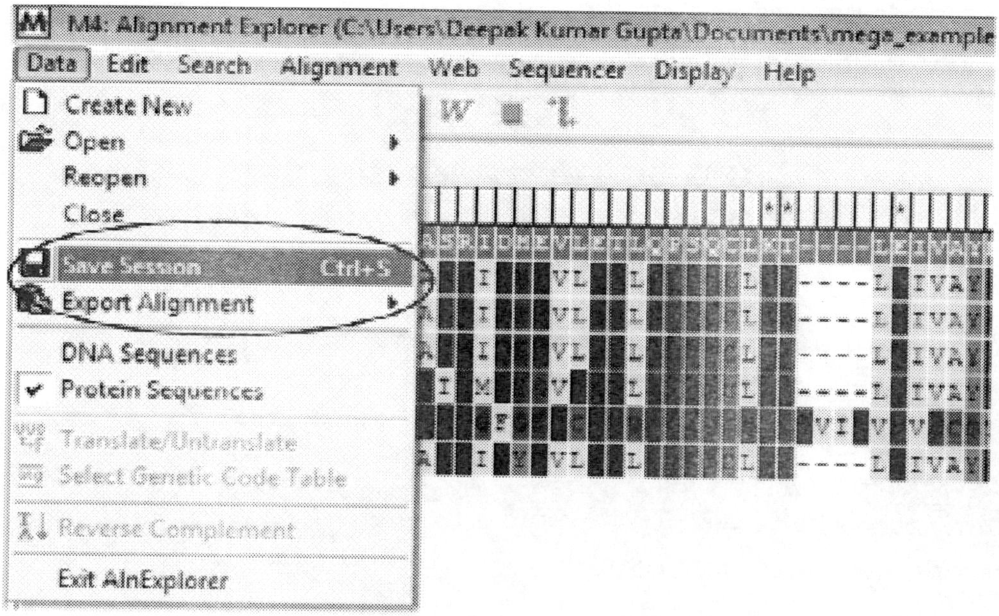

Fig. S8. Saving alignment session.

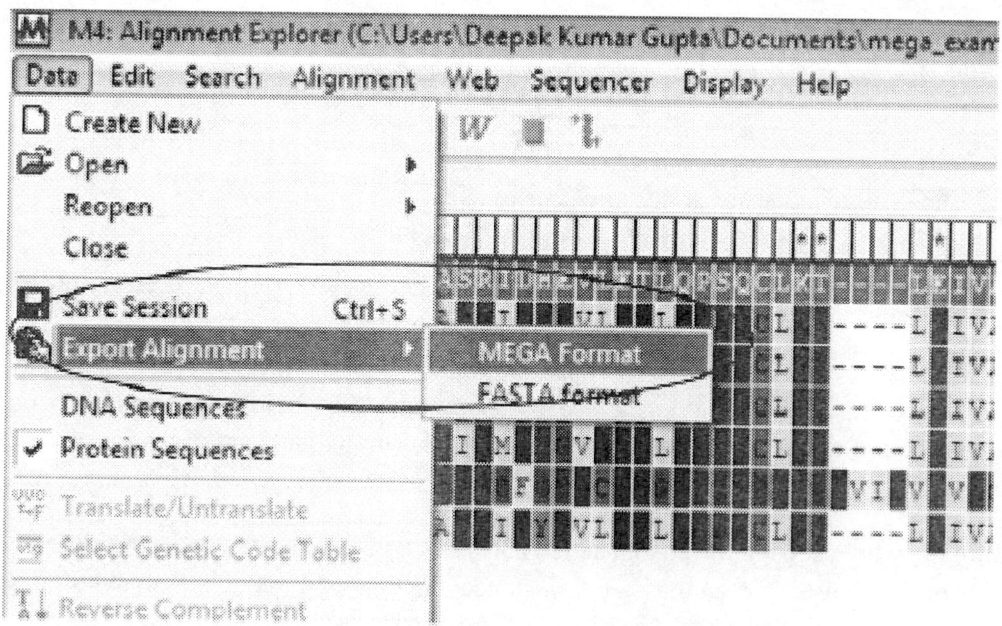

Fig. S9. Saving alignment in MEGA format.

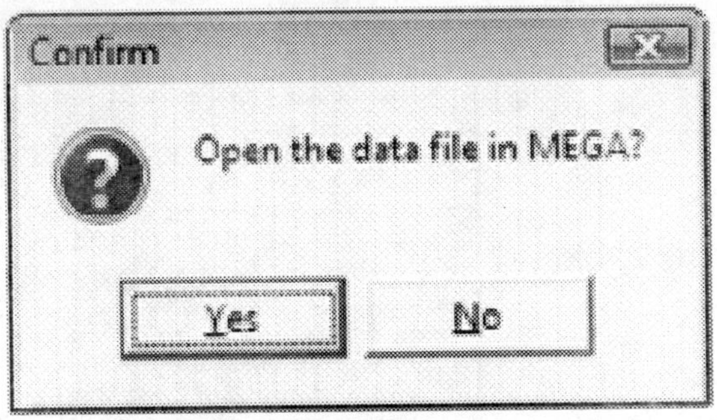

Fig. S10. Confirmation of the data opening in MEGA software.

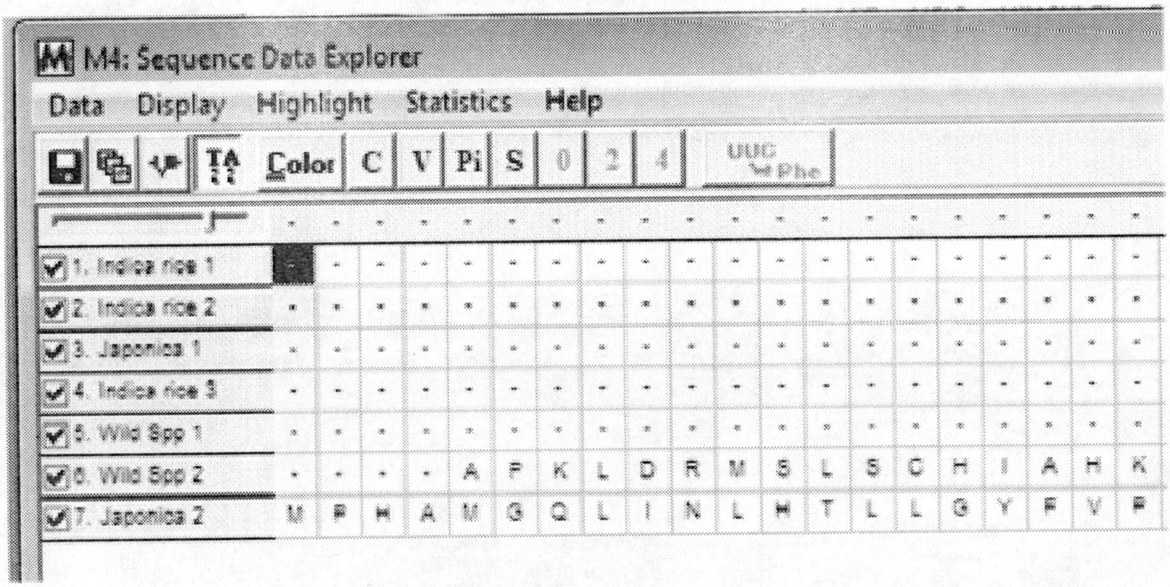

Fig. S11. Data opened in MEGA software.

Step 13. After computing the phylogeny, MEGA generates the phylogenetic tree with bootstrap values given on the nodes of the branches. The names of individuals are given on the termini of the branches. Save the current tree session (Fig. S14).

In this way the whole session of phylogenetic analysis is completed. The results can be interpreted as per the material used for analysis. For instance, in Fig. S14 sequence analysis of specific allele cloned and sequenced from three lines of indica rice, two lines of *japonica* rice and two from the wild species were used for phylogenetic analysis. The alleles isolated from *indica* rice lines formed one cluster and those isolated from *japonica* formed another cluster. However, same allele when isolated from the wild species did not share any of these two clusters. Hence, specific conclusions can be drawn from the output of phylogenetic analysis.

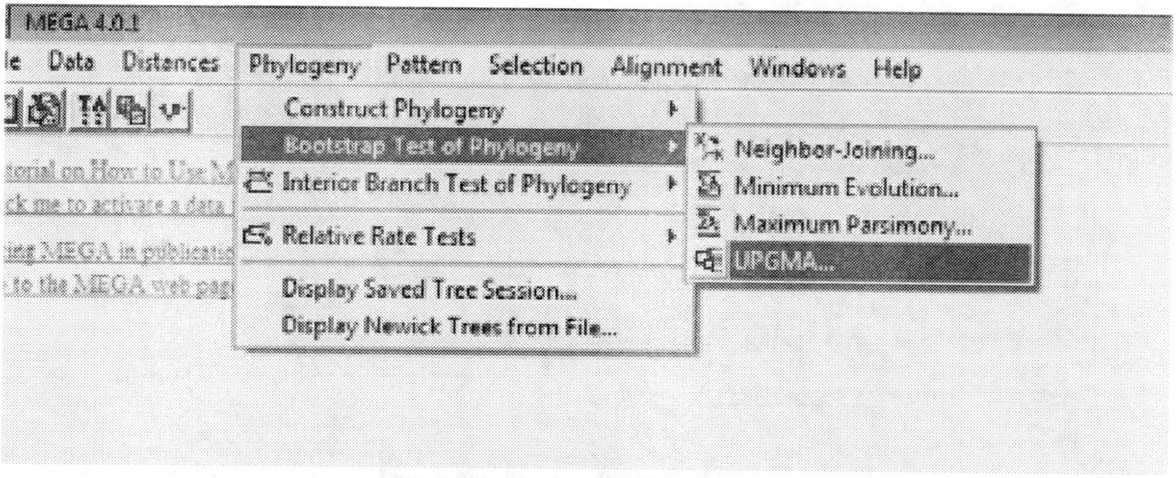

Fig. S12. Selection of methods for phylogenetic analysis.

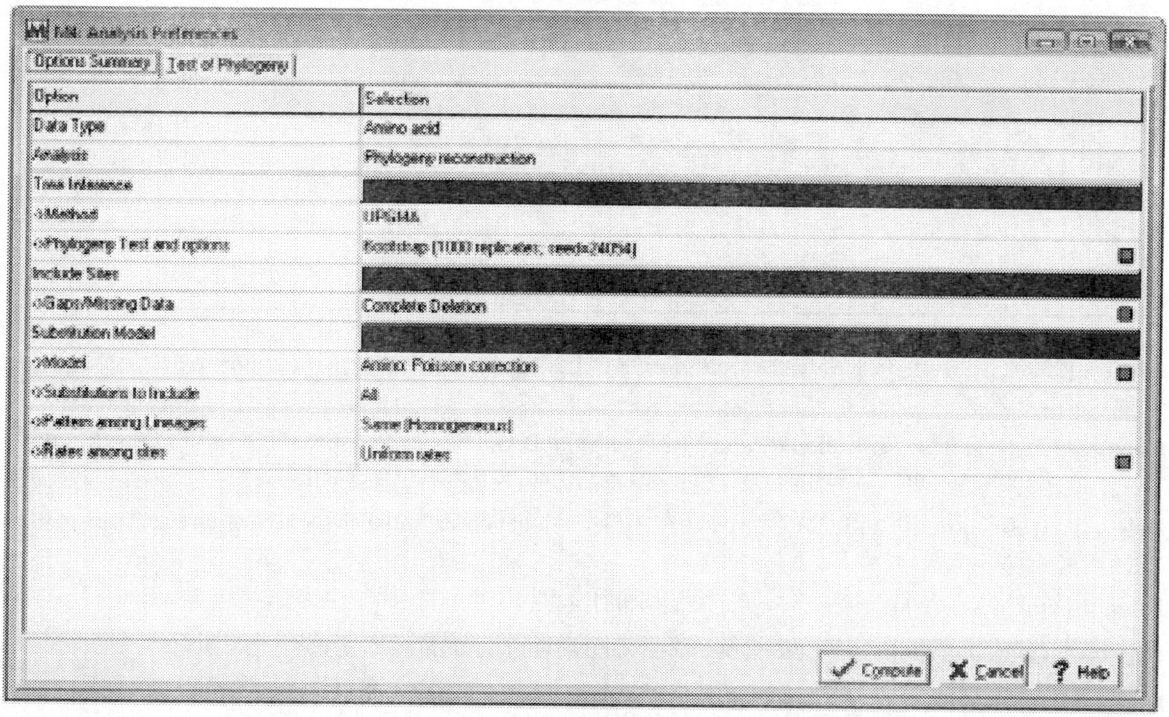

Fig. S13. Parameter selection for phylogenetic analysis.

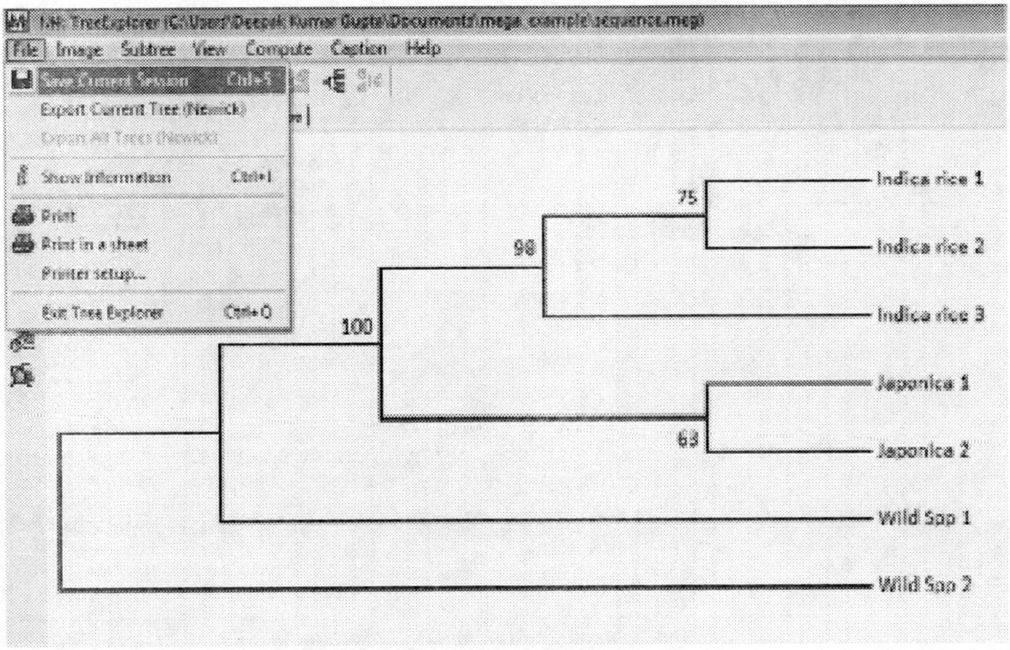

Fig. S14. Phylogenetic tree.

Suggested Readings

Aiyar A (2000), The use of CLUSTAL W and CLUSTAL X for multiple sequence alignment. *Methods of Mol. Biol.* **132**: 221-24.

Chenna, R.H. Sugawara, T. Koike, R. Lopez, T.J. Gibson, D.G. Higgins and J.D. (2003), Thompson Multiple sequence alignment with the Clustal series of programs. *Nucleic Acids Res.* **31**: 3497-3500

Delsus, F., H. Brinkmann and H. Philippe H. (2005), Phylogenomics and the reconstruction of the tree of life. *Nature Rev Genet.* **6**: 361-375.

Eisen JA and M Wu (2002), *Theor Popul Biol.* 61:481-487.

Higgins D.G. (2000), Amino acid-based phylogeny and alignment. *Adv Protein Chem.* **54**: 99-135.

Retief J.D. (2000), Phylogenetic analysis using PHYLIP. *Methods of Mol. Biol.* **132**: 243-258.

9

Gene Prediction and Annotation

Basics of DNA and RNA and various DNA sequencing techniques have already been described in the previous chaperts. However, before using any computational tool for gene prediction, we should be very clear about the basic definitions and structures of typical prokaryotic and eukaryotic genes.

WHAT IS A GENE?

Various hypotheses have been put forth by many workers to define a gene. The classical definition states that a portion of DNA that determines a phenotype is called gene. Later Beadle and Tatum (1940) gave another definition which is known as one gene-one enzyme hypothesis. According to this, a single gene is responsible for the synthesis of an enzyme. This was followed by one-gene-one protein hypothesis. However, the current definition of the gene is that "a piece of DNA sequence (in some cases RNA) which produces functional gene products like RNA and proteins.

Genes can be divided into two different categories:

1. Coding genes i.e. the genes which code for structural proteins and enzymes via messenger RNA.
2. Non-coding genes which codes for structural RNA like transfer RNA (tRNA) and ribosomal RNA (rRNA) etc.

The coding region of the gene consists of open reading frame (ORF), whereas the non-coding regions include, regulatory regions (RNA polymerase and transcription binding sites), introns and polyadenylation (PolyA) sites.

A typical gene should have following parts:

- An open reading frame (ORF) with well defined start (ATG) and stop (TGA,TAA,TAG) codons.
- Promoter sequences (TATA, CAAT box) in the upstream region.
- Poly A tail in the downstream region.
- Splice sites having AG and GT splice signals.

All the computation tools used for gene prediction basically look for the presence or absence of the above mentioned specific sequences of the gene in a long string of nucleotides and then predict a gene. Basic feature of typical prokaryotic and eukaryotic genes are given in Fig.1.

GENE PREDICTION METHODS

Gene prediction is defined as the process of prediction of possible genes in a given DNA sequence by a computer programme based on different structural features of the gene.

The gene-finding strategies can be grouped into three methods.

(i) Content-based methods

These methods rely on the over all, bulk properties of a sequence used in decision-making. The characteristic features of this method include, how often particular codons are used, the periodicity

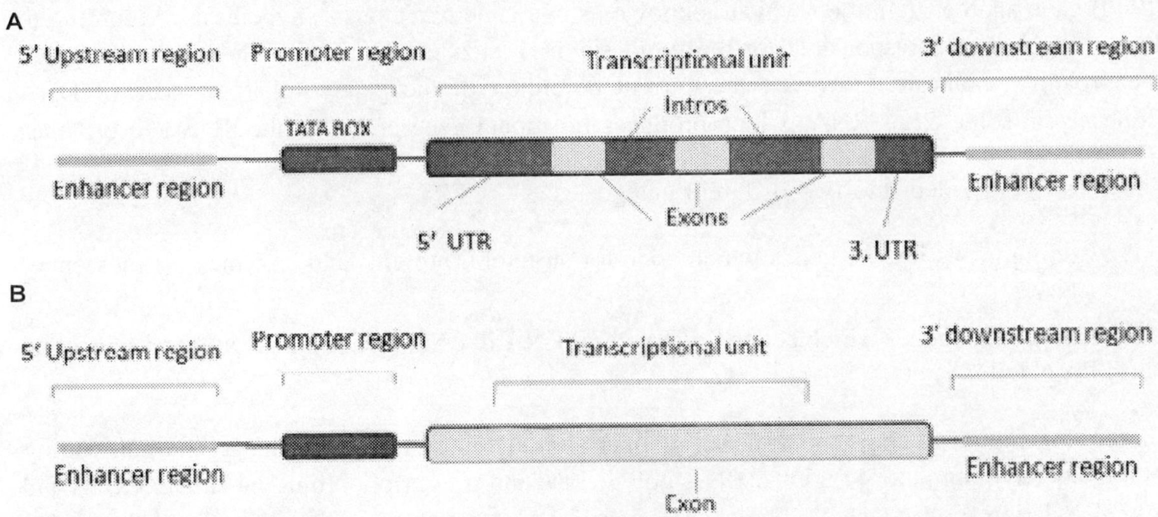

Fig. 1. Typical gene structure showing different parts of a gene. (A) Eukaryotic gene structure. (B) Prokaryotic gene structure.

of repeats, and the compositional complexity of the sequence. Since, the different organisms use synonymous codons with different frequency, such clues can provide insight in determining regions that are more likely to be exons.

(ii) Site-based methods

These methods depend upon the presence or absence of a specific sequence, pattern, or consensus. These are used to detect features such as donor and acceptor splice sites, transcription factors binding sites, polyA tail, and start and stop codons.

(iii) Comparative methods

The comparative methods are based on sequence homology approach. The translated sequences are subjected to database searches against protein sequence in the database to determine whether a region in the query sequence shows significant match with the already characterized genes present in the database.

BIOINFORMATICS TOOLS

Currently, following important bioinformatics tools are used for gene prediction in different genome sequences.

(i) G RAIL

The Gene Recognition and Analysis Inter Link (GRAIL) has been developed by Uberbacher and Mural (1991) and Mural *et al.* (1992). The GRAIL is among the first of the techniques developed for gene prediction which has been used very extensively. The program GRAIL 1 makes use of neural network method to recognize coding potential of a sequence in fixed-length windows of 100bp. It does not take into consideration additional features of genes such as splice sites or start and stop codons. An improved version of GRAIL 1 (called GRAIL 1a) has also been developed which is an expansion of this method. The GRAIL1a considers regions immediately adjacent to the DNA regions deemed to have coding potential. This method results in better performance in both finding true exons and eliminating false positives. Both of these gene finding tools, GRAIL 1 or GRAIL 1a would be appropriate in the context of searching for single exons. A further refinement of the existing method led to the development of a second version, GRAIL 2 (Fig. 2). The revised version considered variable-length widows and contextual information (e.g., splice junctions, start and stop codons, polyA signals) on gene for accurate gene prediction. These software tools are available at http://compbio.ornl.gov/grailexp/.

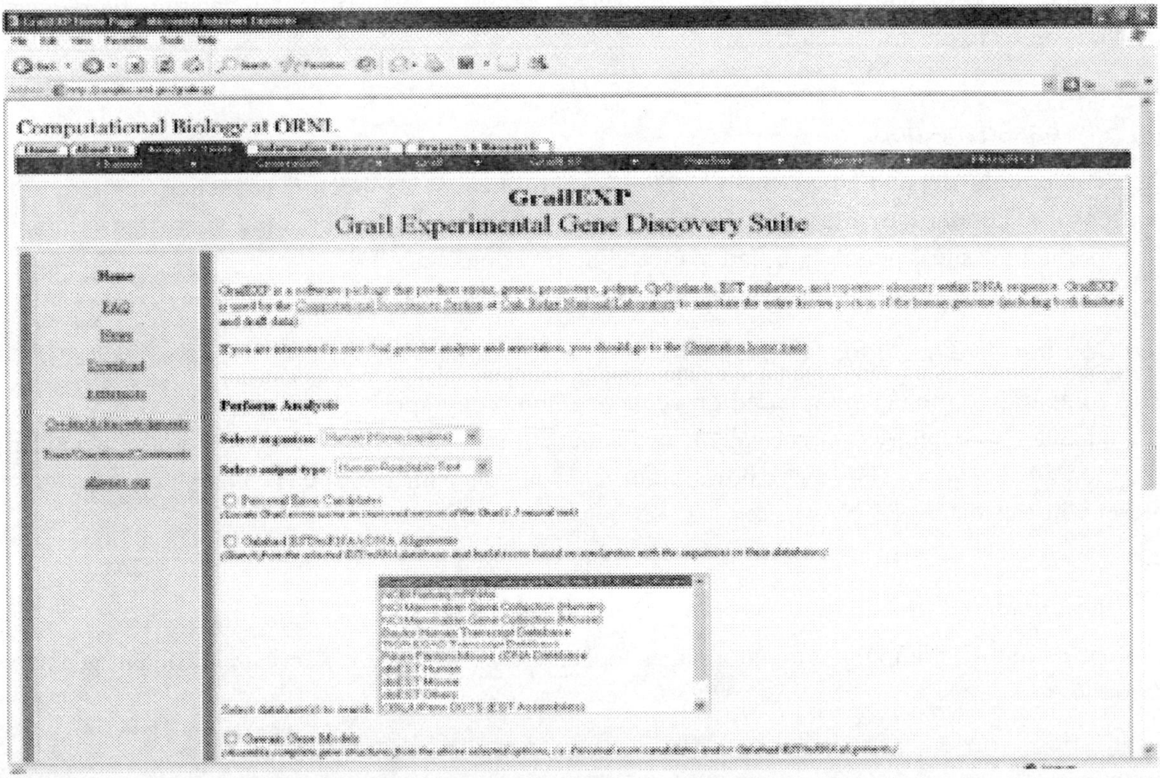

Fig. 2. Input/Query window of GRAIL2.

(*ii*) *FGENESH/FGENES*

This is one of the most popular gene prediction tools used by biologists world over and it has been developed by Victor Solovyev and colleagues (Solovyev *et al.*, 1995) (Fig. 3a). This is a method that predicts internal exons by looking for structural features such as donors and acceptor splice sites, putative coding regions, and intronic region both 5' and 3' to the putative exon Solovyev *et al.*, 1995). This method makes use of linear discriminant analysis, which is a mathematical technique that allows combined analysis of data derived from different experiments. For the combined data analysis, a linear function is used to discriminate between the presence and absence of exons in a given DNA sequence. In FGENESH, results of the linear discriminate approach are then passed to a dynamic programming algorithm, which combines these predicted exons into a best coherent gene model prediction. An extension of FGENESH, called FGENES, can be used in long DNA stretches for multiple gene prediction. Different steps used for gene prediction are given in Fig. 3b.

(*iii*) *GENE ID*

Roderic Guigo and his associates originally developed the Gene ID program in 1992, at the

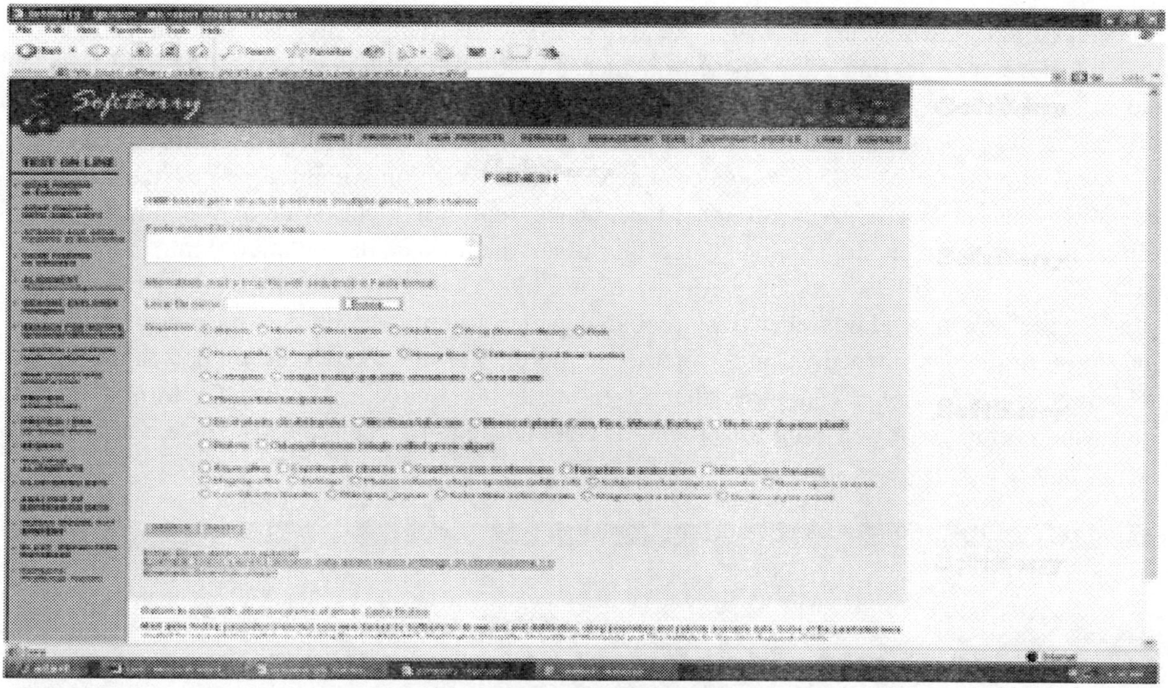

Fig. 3. (a) Input window of FGENESH program.

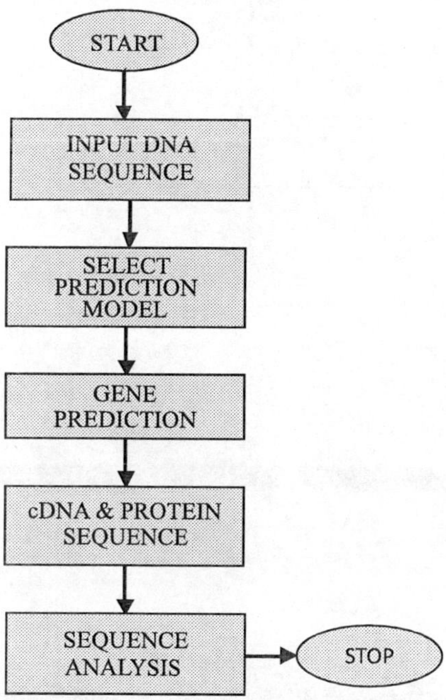

Fig. 3. (b) Process flow chart of gene prediction by FGENESH.

Molecular Biology Computer Research Resource at Harvard University. The current version of Gene ID developed to find exons based on measures of coding potential (Guigo *et al*. 1992). The original version of this program was among the fastest gene prediction programs which used a rule-based system to examine the presence of putative exons and assemble them into the "most likely gene" for that sequence. Gene ID uses position-weight–matrices algorithm to access whether or not a given stretch of sequence represent gene structural features like a splice site or a start or stop codon. Once this assessment is made, putative exon models are built. On the basis of the sets of predicted exons that Gene ID develops, a final refinement round is performed, yielding the most probable gene structure based on the input sequence. Once the program is opened through internet explorer, the main program window appears on the computer screen. One can paste the target nucleotide sequence in FASTA format in the input window and go on selecting the prediction options (Fig. 4). Then, the sequence is submitted for the prediction of gene structure.

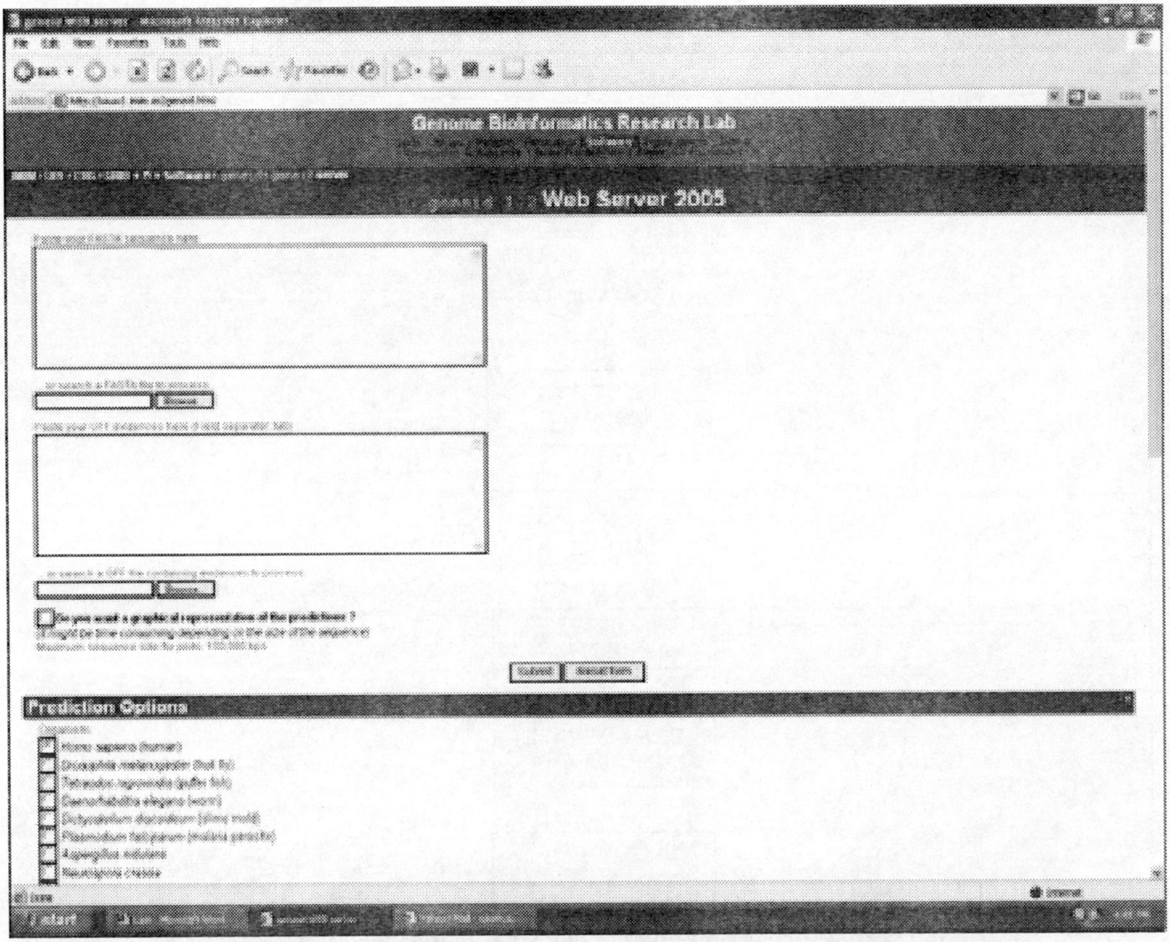

Fig. 4. Input window of GENEID program (http://www1.imim.es/geneid.html).

(iv) GENEPARSER

Gene parser program has been developed by Snyder and Stormo (1993); and Snyder and Stormo (1997). It uses a slightly different approach in identifying putative introns and exons. Instead of predetermining candidate region of interest, GeneParser computes scores on all "subintervals" in a submitted sequence. Once each subinterval is scored, a neural network approach is used to determine whether each subinterval contains the first exons, internal exons, final exons, or intron. The individual predictions are then analyzed for the combination that represents the most likely gene model.

(v) HMM GENE

The HMM gene software, predicts gene structure in any given DNA sequence using a Hidden Markov Model (HMM) method, which is geared towards maximizing the probability of an accurate prediction (Krogh, 1997). The use of HMM in this method of gene prediction helps to access the confidence in any one prediction, enabling HMM gene to report the "best" prediction for the input sequence along with alternative predictions on the same sequence as well. By getting multiple prediction options on the same target region, the user may get an idea of possible occurrence of alternative splicing in the region containing a single gene.

(vi) MZEF

MZEF stands for " Micheal Zhang's Exon Finder," after its author at the Cold Spring Harbor Laboratory, NY, USA. In this software, the prediction relies on a technique called quadratic discriminant analysis (Zhang, 1997). It uses the results of two types of predictions (for instance, splice site scores vs exon length) for plotting against each other on a simple XY graph. If the relationship between these two sets of data is nonlinear or multivariate, the resulting graph will look like a swarm of points. The points lying in only a small part of this swarm will represent a "correct " prediction. A quadratic function is used to separate the correctly predicted points from the incorrectly predicted points. In case of MZEF, the measured variable include exon length, intron- exon strand and frame scores. The software is intended to predict internal coding exons and does not give any other information with respect to gene structure, which is a major drawback of this program.

(vii) GENSCAN

The GENSCAN software, was developed by Chris Burge and Sam Karlin (Burge and Karlin, 1997; Burge and Karlin, 1998). It is designed to predict complete gene structure. As such, GENSCAN can identify different components of a gene like introns, exons, promoter sites and polyA signals, similar to the other gene identification algorithms. Like FGENES, GENSCAN does not expect the input sequence to represent one and only one gene or one and only one exon. It can accurately make prediction for sequences representing either partial gene or multiple genes separated by intergenic regions of the DNA.

GENSCAN relies on the "probabilistic model" (Laplace, 1812, Kolmogrov, 1950) of genomic sequence composition and gene structure. The gene structure descriptions that match or are consistent with the query sequence are identified by the algorithm, then a probability is assigned to a given stretch of sequence which represents an exon and promoter. The "optimal exons" with the highest probability are identified which represents the part of query sequence having the best chance of actually being an exon. This method also predicts "suboptimal exons," stretches of sequences having an acceptable probability value. The users can examine both sets of predictions to identify both alternatively spliced regions of the genes or other non-standard gene structures.

The exons predicted by GENSCAN having a very high probability value (P>0.99) are 97.7% accurate and prediction matches to a true, annotated exon. These high probability prediction can be used in the rational design of PCR primers for cDNA amplification or for PCR based gene cloning in different organisms. The exons predicted by GENSCAN that have probabilities in the range of 0.50 to 0.99 are most of the time deemed to be correct. The best-case accuracies for P-values over 0.90 are on the order of 88%. However, any prediction below 0.50 should be discarded as unreliable. This program is quite popular among biologists. The gene prediction can be performed as shown is Fig. 5.

EXERCISE

Gene Prediction using GENSCAN software from the given sequence data.

GENSCAN is a general-purpose gene identification program which analyzes genomic DNA sequences from a variety of organisms including human, other vertebrates, invertebrates and plants. It has been used here as an example since, it is freely available in the public domain and can be accessed at anytime. It does not involve any license fee etc. Different steps used for gene prediction are given below:

Step 1. Open the GENSCAN web page from http://genes.mit.edu/GENSCAN.html (Fig. S1)

Step 2. Paste, input DNA sequence in the window .

Step 3. Select the organism, print options, sequence name etc. as shown in Fig S1.

Step 4. Click on the **RUN GENSCAN** to get the output of the program.

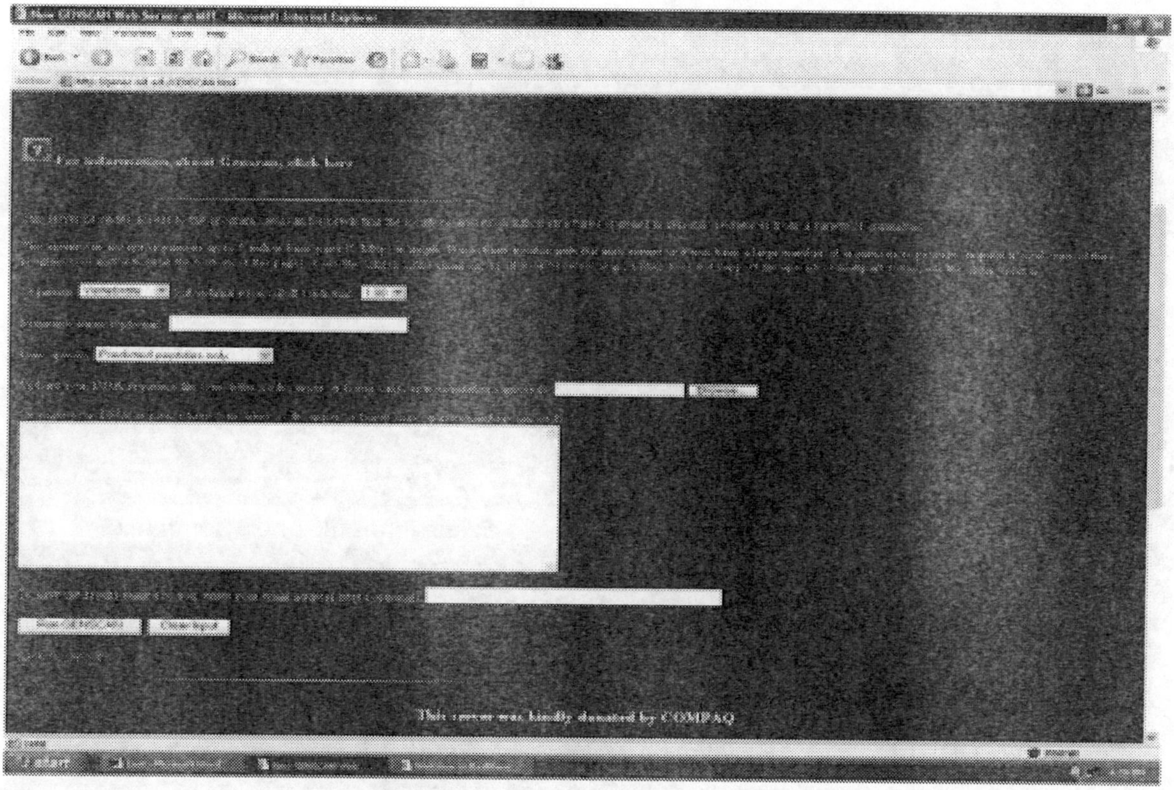

Fig. 5. GenScan program input window.

Step 5. An output of the results page will appear within few seconds (Fig. S2). This will contain summary of the results, protein and mRNA sequences of the predicted genes. The abbreviations used in the results are also explained in the results output window.

GENE ANNOTATION

Once gene prediction is over, we are always curious to know what these genes are doing in a living organism. What trait is controlled by this gene? Therefore, a note on the description of a gene about its function is known as gene annotation.

Gene annotation can be of two types.

Structural gene annotation

In structural gene annotation, one can look for the number of ORFs and size, intrinsic regions, presence of regulatory regions and size as well as GC contents of the genes etc.

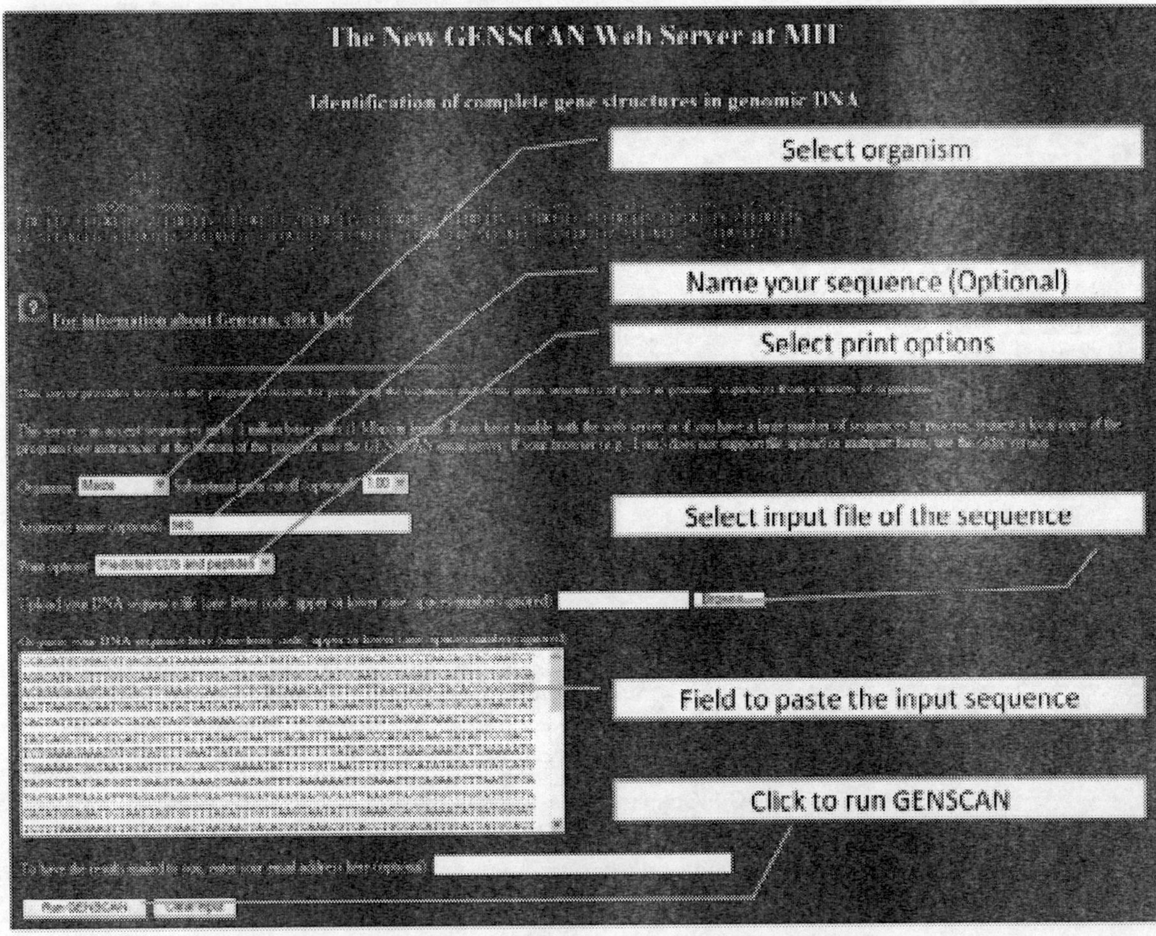

Fig. S1. Steps to be followed in Gene prediction tool.

Functional gene annotation

Once genes are identified, their biological functions have to be determined. Computationally, it can be better performed by using BLSAT P tool against a curated protein database like Swissprot etc. BLAST results are looked carefully and based on the significant matches available in the databases, the biological or biochemical functions, gene expression and involvement of genes in regulation, interaction and expression etc. A particular note is attached to the *in silico* identified genes.

Gene annotation is basically performed by using BLAST tool and finding similarity to the unknown gene in the databases. Based on the similarity the genes are divided into different categories as described in Table 1.

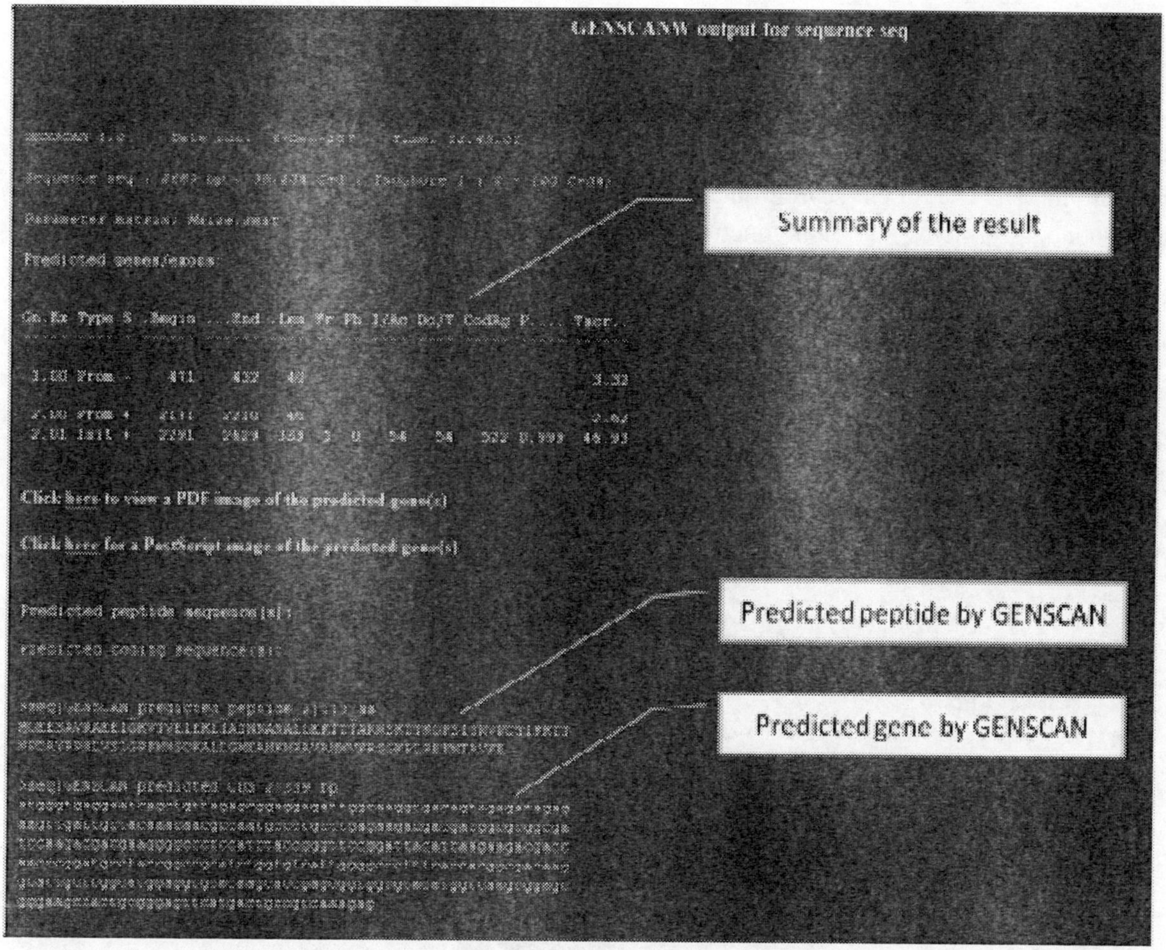

Fig. S2. An output window of Genscan software.

Table 1. Gene annotation standards used in large genome sequencing project.

Gene Category	Description of the match
Known gene	Sequence with 100% identity at the amino acid level to known proteins
Putative gene	Sequence with less than 100% identity but with significant homology to known proteins
Unknown gene	Sequence with homology to unknown ESTs
Hypothetical genes	Sequence predicted by multiple gene prediction programs with no homology to an EST

EXERCISE

Gene annotation by using Bioinformatics tools from the sequence of the predicted gene.

We can perform gene annotation of any DNA sequence of a gene by using BLASTX programme. BLASTX will search query gene sequence against protein database to find out the corresponding proteins and their functions.

Step 1. Open BLAST page of NCBI. (www.ncbi.nlm.nih.gov/blast/Blast.cgi) Select **BLASTX option** from the Basic Blast menu (Fig. S1).

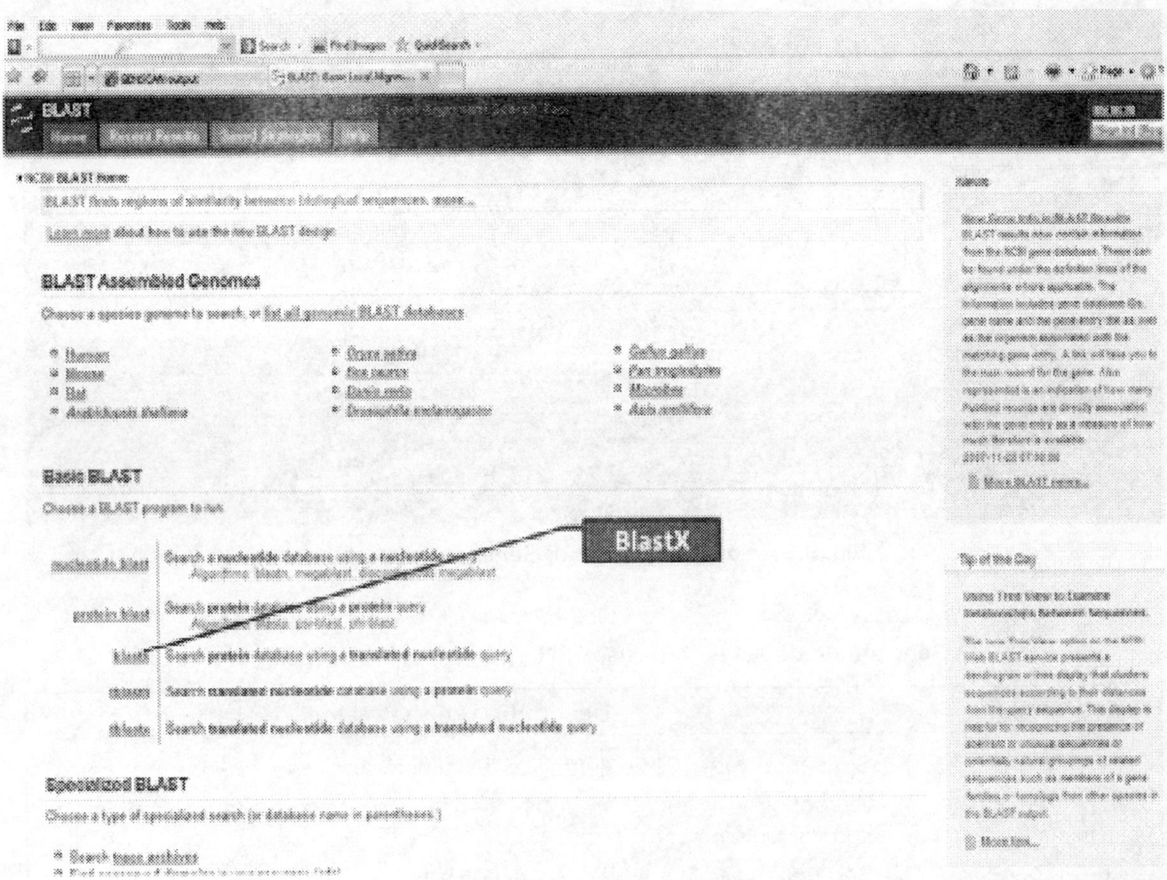

Fig. Sa1. Home page of BLAST.

Step 2. Paste query sequence in the field provided in the window, select the database as **nr**, and set other parameters to make the search specific and click on BLAST to perform the search (Fig. S2).

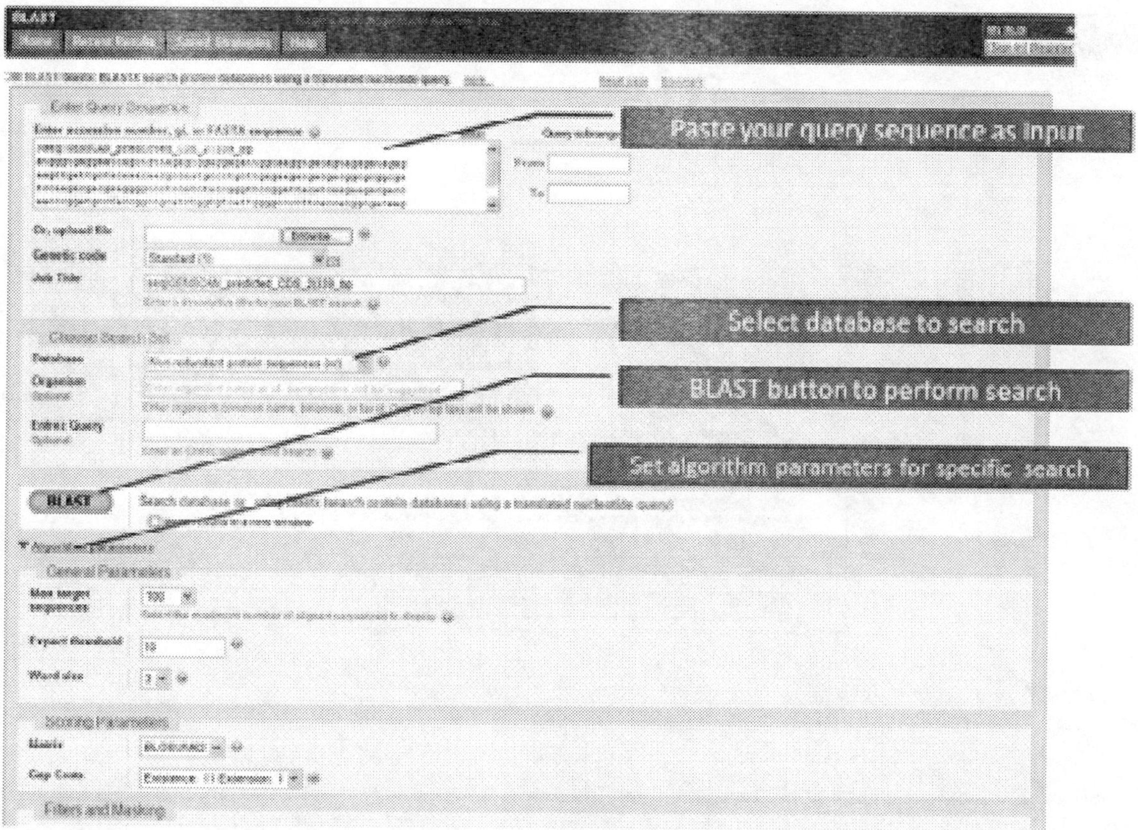

Step 3. In the BLAST X result page click, on the HITs to get the annotations of the genes in the database where query sequence showed hits (Fig. S3).

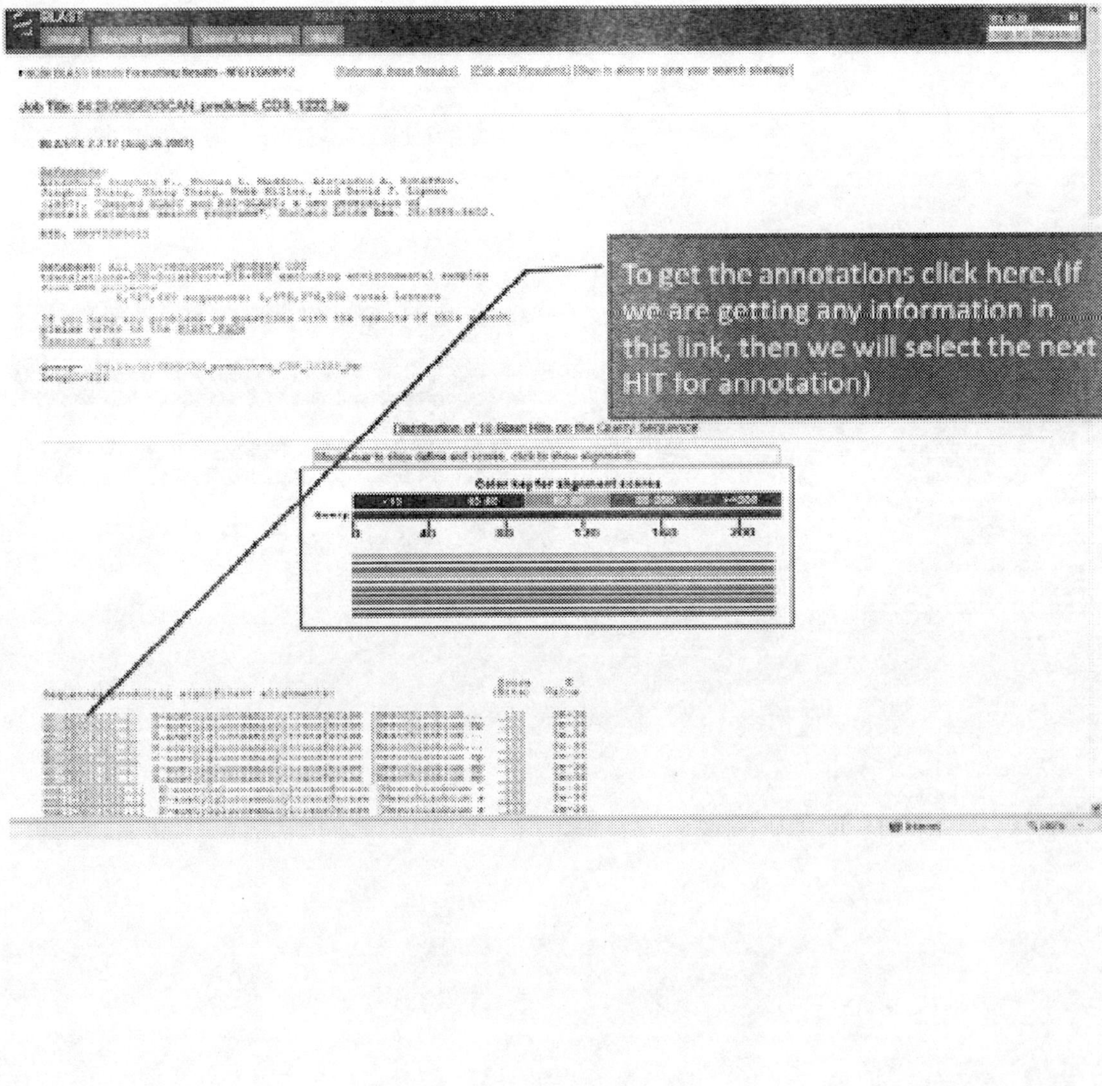

Step 4. The annotations of the targeted gene can be obtained from the results (Fig. S4). It shows that the sequence which we used for gene prediction and annotation code for a protein known as N-acetyleglucosaminyltransferase. Hence, to an unknown sequence we are able to put annotations.

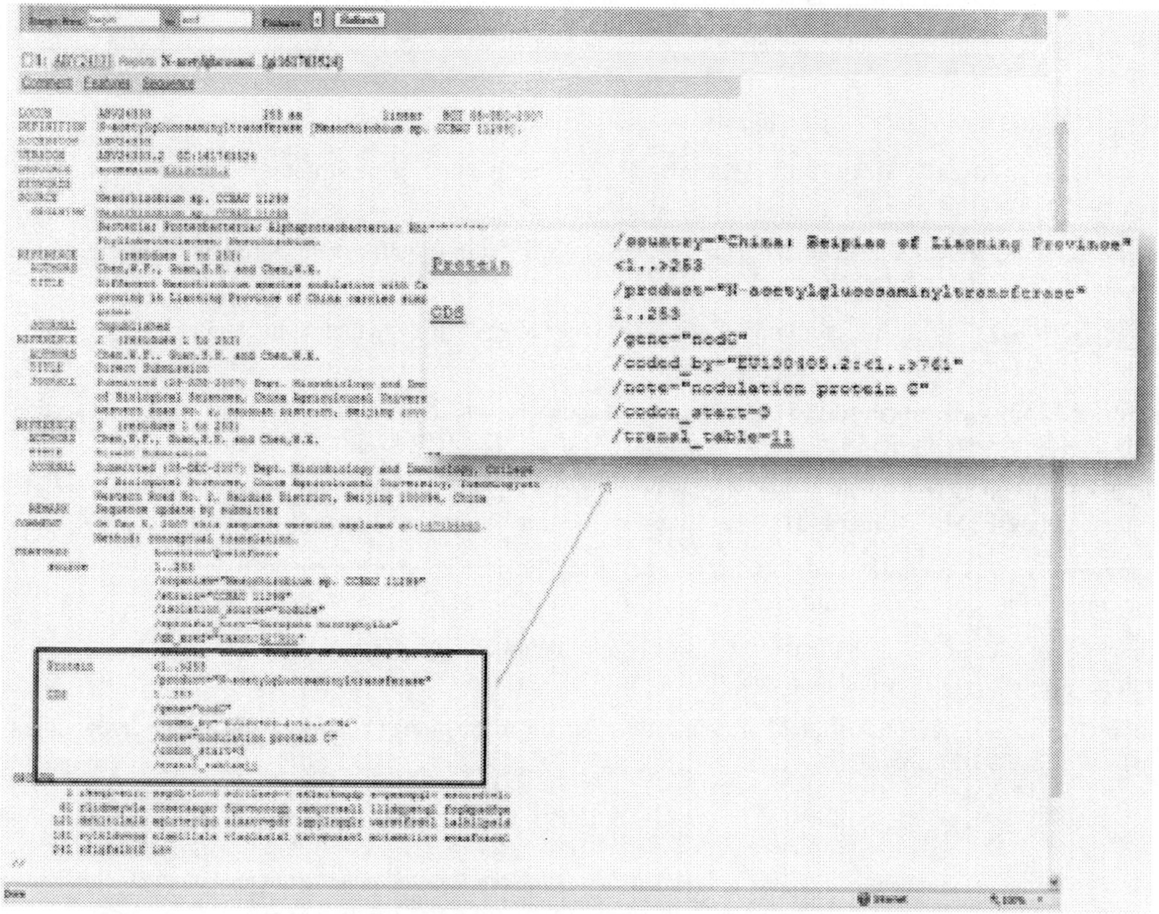

Fig. S4. Results page of the annotations.

Suggested Readings

Baxevanis A.D. (2001), Predictive methods using DNA sequences. In Bioinformatics: a practical guide to the analysis of genes and proteins, eds. A.D. Baxevanis and B.F.F. Ouellette (John Wiley & sons) pp 233-252.

Burge C. and Karlin S. (1997), Prediction of complete gene structures in human genomic DNA. *J. Mol. Biol.* 268, 78-94.

Burge C.B. and S Karlin (1998), Finding the genes in genomic DNA. *Curr Opin. Struct. Biol.* **8**: 346-354.

Guigo R, S Knudsen, N Drake and T.F. Smith (1992), Prediction of gene structure. *Journal of molecular biology*, **226**: 141-157.

Guigo R. (1998), Assembling genes from predicted exons in linear time with dynamic programming. *Journal of computational biology*, **5**: 681-702

Krogh A. (1997), Two methods for improving performance of an HMM and their application for gene finding. *Proc. Int. Conf. Intell. Syst. Mol. Biol.* **5**: 179-186.

Snyder E.E. and G.D. Stormo (1993), Identification of coding regions in genomic DNA sequences: an application of dynamic programming and neural networks. *Nucl. Acid Res.* **21**: 607-613.

Snyder E.E. and G.D. Stormo (1997), Identifying genes in genomic DNA sequences. In DNA and protein sequence analysis, M.J. Bishop and C.J. Rwalings eds. (Oxford University Press) pp 209-224.

Solovyev V.V., A.A. Salamov and C.B. Lawrence (1995), Identification of human gene structure using linear discriminant functions and dynamic programming.*Proc Intl Conf Intell Sys Mol Biol.* **3**: 367-375.

Uberbacher E.C. and R.J. Mural (1991), Locating protein-coding regions in human DNA sequences by a multiple sensor-neural network approach. *PNAS.* **88**: 11261-11265.

Zhang M.Q. (1997), Identification of protein coding regions in the human genome by quadratic discriminant analysis. *PNAS.* **94(2)**: 565-68.

10

DNA Marker Data Analysis

Variation at genetic level can be detected in the form of phenotypic expression of different characters (traits) under varied environment. These identifiable traits are called markers. Transmission of characters from parents to progenies, through genetic material and expression of those differentiating characters in the progenies are known as genetic markers. Before we use DNA markers data analysis tools, it is imperative to understand different types of markers. It is also important to know the basis of different marker systems and how molecular marker data are generated by the biologists.

DIFFERENT TYPES OF GENETIC MARKERS

1. **Morphological markers**: Different plant characters like plant height, leaf shape and pathogen characteristics like culture growth, pigmentation, pathogenecity and other visual identifiable characters are known as morphological markers.

2. **Biochemical markers**: These include analysis of proteins and enzymes like isozymes or allozymes extracted from the tissues. Proteins and isozymes are used as a marker after distinguishing the polymorphisms in the electrophoretically separated and specifically stained protein bands.

3. **DNA markers**: Analysis of DNA in the form of individual bands or restriction fragments is the direct method of estimating large number of differences at genetic level.

Morphological and biochemical markers are the ultimate result of gene expression, which are influenced by the environment, as well as developmental stages of the organism under study.

Besides, these markers are a few in numbers and do not represent the whole genome of an organism hence cannot be used for the construction of a saturated genetic map. However, DNA markers are enormous in numbers and stably inherited in Mendelian fashion. These are also not affected by environment or developmental stages of the plants. Variation at DNA level is detected by DNA polymorphisms survey by various molecular techniques. Polymorphisms at the DNA level is the result of simple point mutation, insertion or deletion of DNA segments etc.

TYPES OF DNA MARKERS

Natural variation at DNA level can be detected by a range of molecular biology techniques developed in the recent past. The choice of their use solely depends on nature of the study and species in question. Different DNA markers have been classified into three categories on the basis of molecular biology techniques used.

Hybridization-based markers

The most important hybridization based techniques is Restriction Fragment Length Polymorphism (RFLP). In this technique, DNA is first digested with restriction enzymes, separated on an agarose gel and Southern transferred to membrane filter. DNA fixed on the filters is then hybridized with locus specific DNA probes labelled with radioactive or non-radioactive substance and DNA fingerprints are obtained by autoradiography. Alternatively, hybridization can be performed with microsatellite specific DNA probes. Such type of DNA markers are then known as Variable Number Tandem Repeats (VNTR) and oligonucleotide fingerprinting. Anonymous cloned or PCR amplified DNA sequences may also be used as probes to detect variation.

Sequence targeted and single-locus PCR- based markers

Though DNA sequence information can be rapidly detected by the use of DNA markers, it provides incomplete information. Moreover, some of the PCR-based DNA markers do not provide allelic information. Such problems can be overcome by the use of PCR primers flanking to a specific locus. Most ecological and evolutionary studies are based on DNA markers derived from the sequences of DNA. However, it is a labour intensive and expensive technique and sometimes difficult to use in the detection of polymorphisms at species or race level. For the specific and sensitive detection of locus, PCR techniques like, allele specific Oligonucleotide Ligation Assay (OLA) and after cloning and sequencing of simple sequence repeats, PCR primers flanking to the region can be designed. These are known as Sequence Tagged Microsatellite Site (STMS) markers.

PCR-based markers

Amplification based DNA markers have been proved versatile and easy because there is no need of difficult and lengthy steps of Southern blotting. All the genetic markers based on PCR are generated by the modifications of original method of PCR. The technique is very sensitive and allows specific amplification of DNA fragments from the genomic DNA of all the organisms. PCR is a simple chemical synthesis of DNA molecule with the help of DNA polymerase and primers by using genomic DNA as template. The technique can be automated for large-scale applications and devoid of radioactivity used for labelling DNA probes. The application of PCR technology is almost countless. Brief description of important PCR- based markers is given below:

PCR-RFLP markers

The basic principle of PCR-RFLP techniques is exactly similar to normal RFLP analysis with a distinction that no probe is used here for hybridization. In this technique, two primers are used to amplify a specific region of the genome. The amplified fragment is then digested with restriction endonucleases, separated on agarose/polyacrylamide gel to detect polymorphic DNA profiles. The obvious advantages of PCR-RFLP are its speed and sensitivity to detect variation. The whole experiment can be performed within 24 hours, with small amount of genomic DNA compared to RFLP analysis, which takes about 24 hrs and require 4-5 μg DNA per sample.

Random amplified polymorphic DNA markers

One of the important molecular markers developed by the use of PCR technique is Random amplified polymorphic DNA (RAPD). This technique is a variant of PCR where a single primer developed from randomly chosen ten nucleotides are used for amplification of the templates. After DNA denaturation in PCR, the primers randomly join with the homologous DNA sequences at the template. Since, primers are short in length and having random homologous DNA sequences, low primer annealing temperature (35-37°C) is used to ensure perfect primer annealing to the template. Various steps of RAPD are given in Fig.1. DNA fragments are amplified randomly wherever primers anneal to the opposite strands facing their 3′ end. PCR products are separated on an agarose gel, stained with ethidium bromide and visualized under UV light (Fig.1). RAPD are being used in various molecular biology studies. It has several advantages over the existing RFLP technique. It is very rapid and easy technique where no sequence information is required for generating primes. This involves no radioactivity, is less expensive and does not require rightys killed manpower.

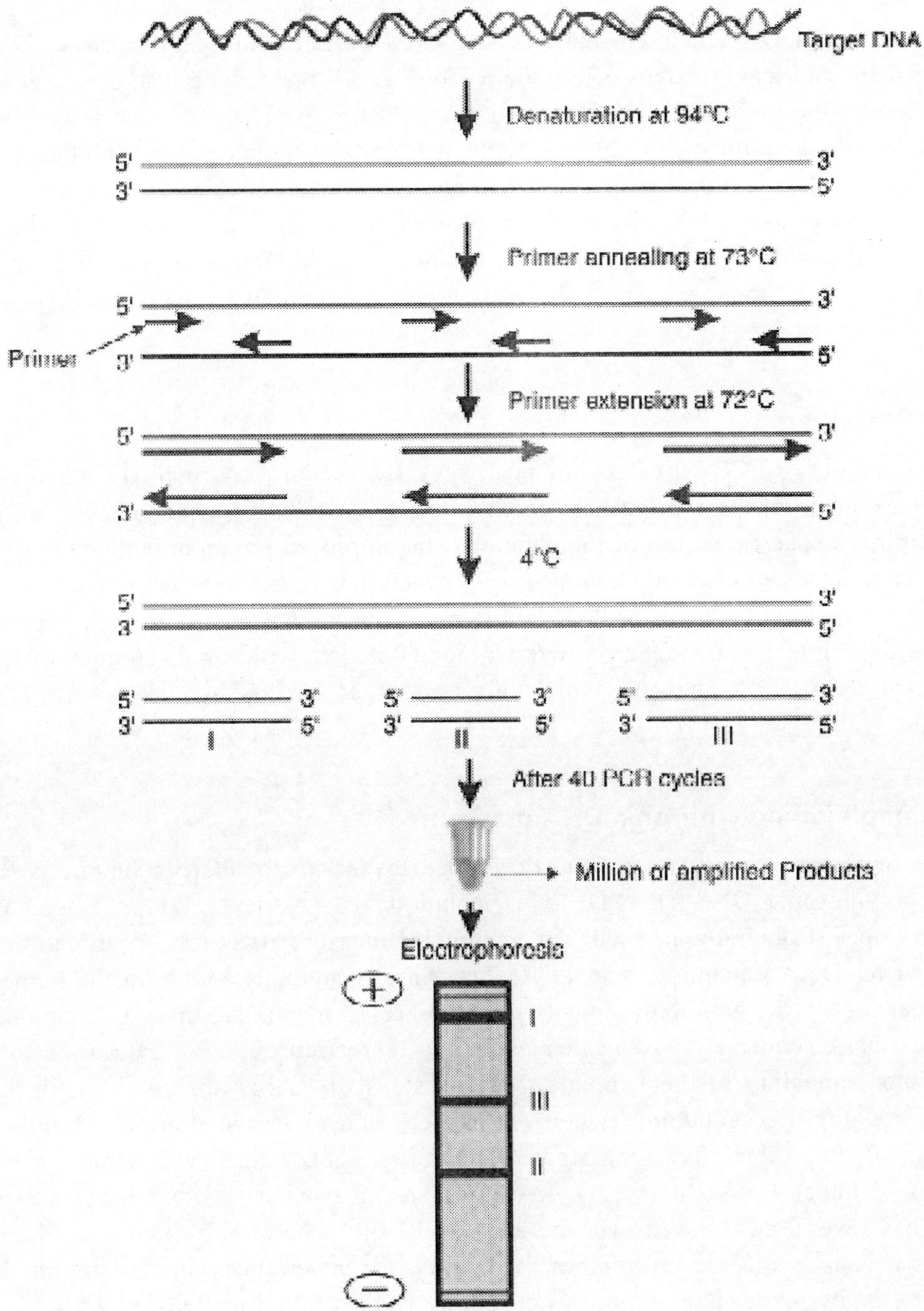

Fig. 1. Basic steps involved in RAPD analysis. Upon electrophoresis, PCR products are seperated according to their size.

DNA amplification fingerprinting markers

DNA amplification fingerprinting (DAF) is another PCR technique in which a short arbitrary primer (5-6 nucleotide in length) is used for PCR to obtain reproducible banding patterns on agarose or acrylamide gels. Like RAPD, DAF does not rely on previous knowledge of genome variation. It is a simple, fast and sensitive method that may be used in a wide variety of organisms.

Amplified fragment length polymorphism markers

Amplified fragment length polymorphism (AFLP) is a good combination of RFLP and PCR. In this technique, genomic DNA is first cut with two restriction enzymes simultaneously, one of which is a rare cutter (*Eco* R1) and another one is a frequent cutter (*MSe* 1). At both the cut ends, DNA adapters with known DNA sequences are ligated. The base sequences of these adaptors are used for designing primers for the PCR amplification. Besides, while synthesizing primers, 1-3 additional known base pairs are added at the ends of the primers. Hence, PCR amplifications obtained where the primers anneal to fragments, have the adaptor sequences plus the complementary base pairs to the additional nucleotides. These additional base pairs are thus known as selective nucleotides. The purpose of adding selective bases is that only fewer restriction fragments ranging from 50-100 bp are amplified wherever the complementary primers are annealed to the adaptors at the cut ends. After PCR, the amplified fragments are separated by electrophoresis on a polyacrylamide sequencing gel. Fingerprints are normally obtained after exposing an x-ray film on the gel, if one of the PCR primers is radio-labelled with P^{32}. However, non-radioactive detection by silver staining of the polyacrylamide gel is also done to get AFLP fingerprints. Different steps involved in AFLP techniques are given in Fig. 2.

Microsatellite markers

Microsatellites are the simple sequence tandem short oligonucleatide repeats distributed throughout the genome. The PCR based method, can be used to detect co-dominant polymorphisms at simple sequence repeats (SSR) loci. Microsatellite loci are highly reproducible and generate large number of polymorphic banding patterns upon electrophoresis. Amount of genetic variation in a particular species is mainly dependent on differences in the number of randomly repeating units at a locus. Short oligonucleotides SSR are used as hybridization probes in Southern blotting to generate highly variable DNA fingerprints of the species. Alternatively, PCR primers flanking to the SSR can be developed and variation can be detected on the basis of PCR amplification, separating amplified products on Poly Acrylamide gel Electrophoresis (PAGE) and fingerprints can be obtained by silver staining of the gels. However, SSR-primer development cost is quite high since it requires determination of base sequences of the DNA regions flanking to the SSR. Besides, synthesis of oligonucleotide primers is expensive, though running cost of assay is less

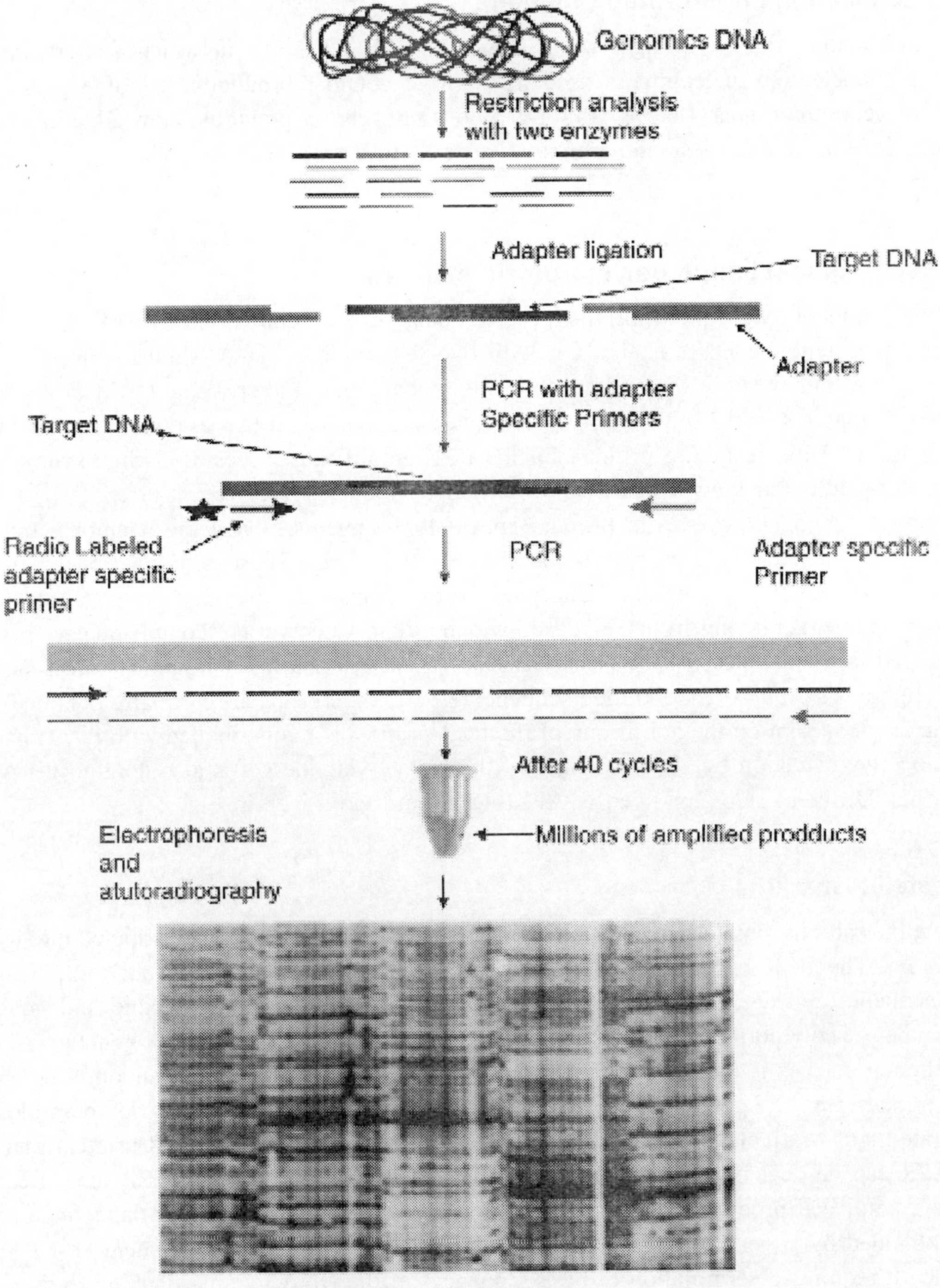

Fig. 2. Steps involved in AFLP analysis.

and is very simple to perform. Different steps used in microsatellite based analysis are given in Fig. 3.

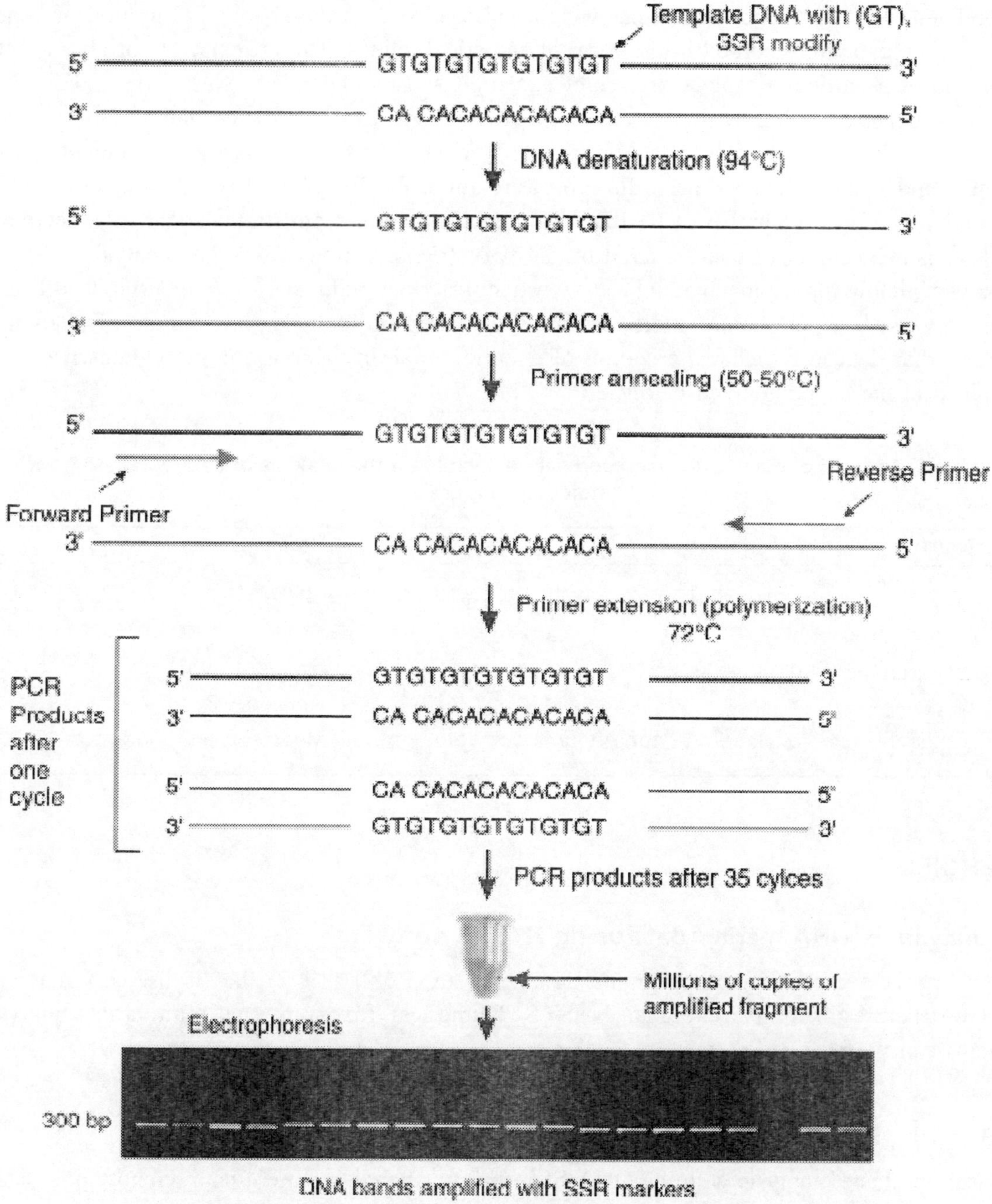

Fig. 3. Basic steps involved in SSR marker analysis.

Computational Analysis of Molecular Data

The molecular data obtained on the gels or autoradiographs is in the form of banding patterns. These bands are of different molecular weight sizes. Based on the size and migration of bands on a gel, the data are recorded in the form of statistical values. The presence '1' or absence '0' of a band of a particular molecular weight is scored as two alleles at a single locus. This data is used to construct a binary matrix on an excel sheet. An example of the binary data is given in Fig. S1. The DNA marker data are analyzed on the basis of similarity co-efficient (F) of strains x and y and calculated according to the relation, F= 2Nxy/(Nx+Ny) (Nei and Li, 1979). Where, Nx and Ny are number of RAPD/RFLP bands obtained from strains x and y, respectively and Nxy is the number of bands shared by the two strains. Various DNA data analysis software are now available for phylogeny studies (www.ucmp.Berkeley.edu/ subway/ phylo/phylosoft) and commonly used in population studies of fungi are listed in Table 1. An example of analyzing DNA marker data in the characterization of genetic variability among the individuals has been explained in the exercise given below.

Table 1. Sources of commonly used software available for the analysis of data obtained from molecular markers

Software	Website	Reference
Phylip	//evolution.genetics.Washington.edu	Felsenstein,1993
Winboot	//www.irri.org	Yap and Nilson, 1996
PAUP	//paup.csit.fsu.edu	Swofford, 2001
NT-SYS	–	Rohlf, 1993
MacClade	//phylogeny.arizona.edu/macclade/	Maddison and Maddison,1992

EXERCISE

A. Analysis of DNA marker data using NTSYS software

NTSYS-pc is a collection of programs developed by Rohlf (1992) that is used to find and display structure in multivariate data. NTSYS (Numerical Taxonomy and multivariate analysis System) can perform many types of analysis.

NTSYS File Format

To perform cluster analysis with the DNA fingerprint or banding pattern data (Genotyping data), prepare excel file for input to the program. NTSYS files are in the form of "matrices" of "1" i.e. presence of band of a particular molecular weight in one individual and "0" i.e. absence of DNA band in another individual.

Preparation of input data

The input data file is prepared in an excel sheet to score genotypic data as instructed and shown in Fig. S1. The description of this file format is as follows:

- To facilitate computer analysis, the presence of a band should be scored as '1', absence as '0' and missing data as 9, in the excel sheet.
- The number in first cell is a code for the type of matrix. The matrix codes are:

  1 = rectangular data matrix
  2 = symmetric dissimilarity matrix
  3 = symmetric similarity matrix
  5 = tree matrix for dissimilarity data
  6 = tree matrix for similarity data

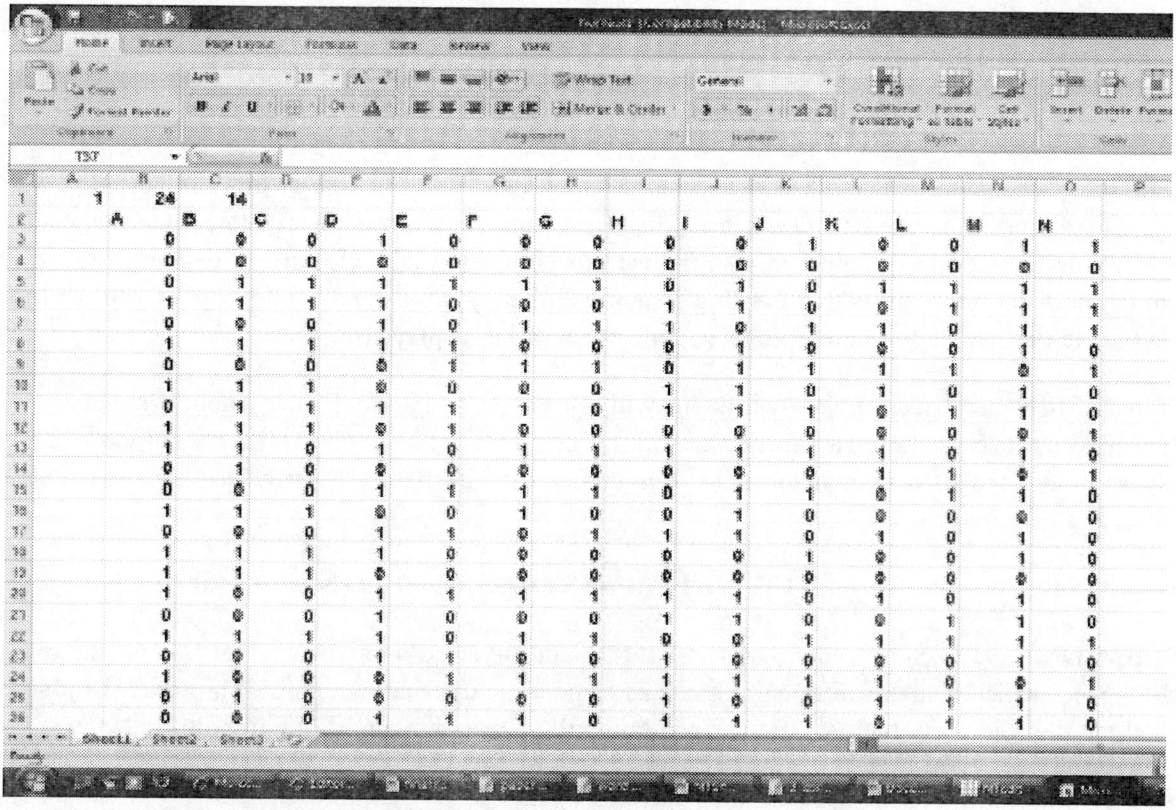

Fig. S1. Input data file of fingerprints or marker data in excel file format.

- The second and third numbers are the numbers of rows and columns in the matrix
- The fourth number is 0, if there are no missing data and 9, if some data is missing in the matrix. The missing data cannot be represented as a blank in the data matrix.
- Second row include the name of samples used for analysis and column includes the marker (Do not write names of marker).

For example we have taken a set of data consisting of 14 individual samples as A to N (Fig. S1). The first column and first row of the excel sheet is represented as 1(rectangular data matrix). The first row and second column is represented as 24 (no. of rows in the matrix) and the first row and third column is represented as 14 (no. of columns (individual markers) in the matrix).

(Note: some time 'o' alphabet is written instead of '0' (zero) by mistake as both the keys are near to each other on keyboard, which will create error during analysis)

- Now save the excel file in NTSYS-pc folder.

RUNNING NTSYS

Step 1. Open the NTSYS-pc folder, double click ntedit.exe. icon.

A window as shown in Fig. S2 will appear in which data from the excel file is imported. The ntedit.exe window is used to check the file. This window is used for creating or editing an NTSYS-pc matrix file. NTedit can also be used to view and make changes in existing files. A limitation of the program is that existing files must already be in a proper format. If you try to load a file that is not formatted properly, an error will be displayed.

Step 2. Click on FILE, a pop-down menu will appear, click import Excel. Another menu will appear which takes data in two formats. Click on using OLE as shown in Fig. S3. Import excel file and save the data with new name. Remember the file name. Now close this ntedit.exe window.

Step 3. Save the data file as NTSYS files in NTSYS-pc folder, as shown in Fig. S4.

Step 4. Open NTSYS-pc folder again; click the icon ntys.exe as given in Fig. S5. A window will appear which contains different buttons as General, Clustering, Graphics, Similarity. Double Click on the Similarity button. Another window will appear showing different buttons.

Step 5. Click on the SimQual button as shown in Fig. S6.

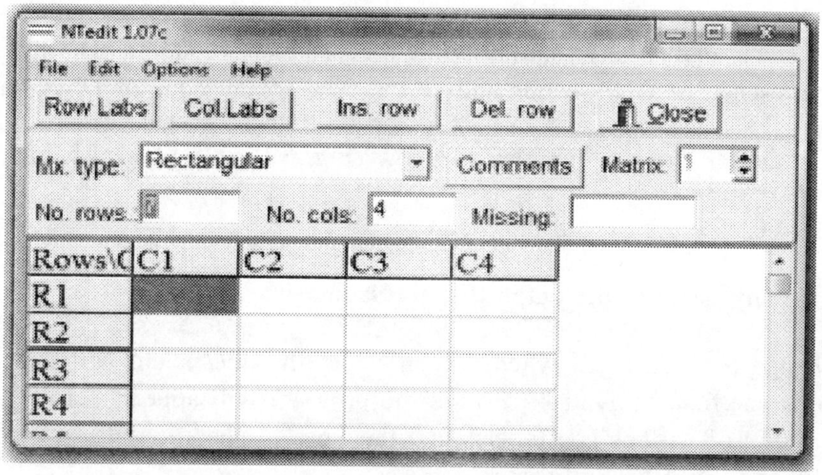

Fig. S2. NT edit folder for importing data from excel sheet.

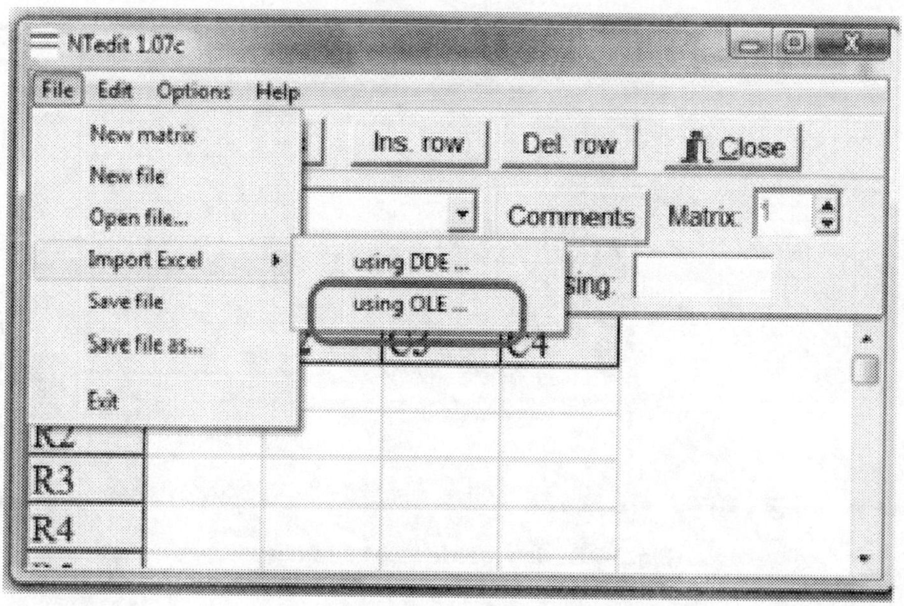

Fig. S3. Importing data in NTedit file.

SimQual computes a variety of similarity and distance coefficients for qualitative data. This program can be used to compute the similarity matrix for binary (banding pattern) data, in which "1" is used for presence of DNA bands and "0" is used for absence of DNA bands.

Step 6. Enter the data in the simQual data window (Fig. S7) as explained below:

Input file-The name of file saved in the ntedit.

Output file- Give any name to the output file for saving it.

Coefficient- Change to J (jaccard) when you move to the coefficient entry, a pop-up menu giving the various coefficients available within the program will appear, Scroll down to J and select it. (Description of various coefficients and their use in specific data type is given in the help file available with NTSYS software).

Then, click to compute button as shown in Fig. S7. A report file will appear (Fig. S8). Close the window.

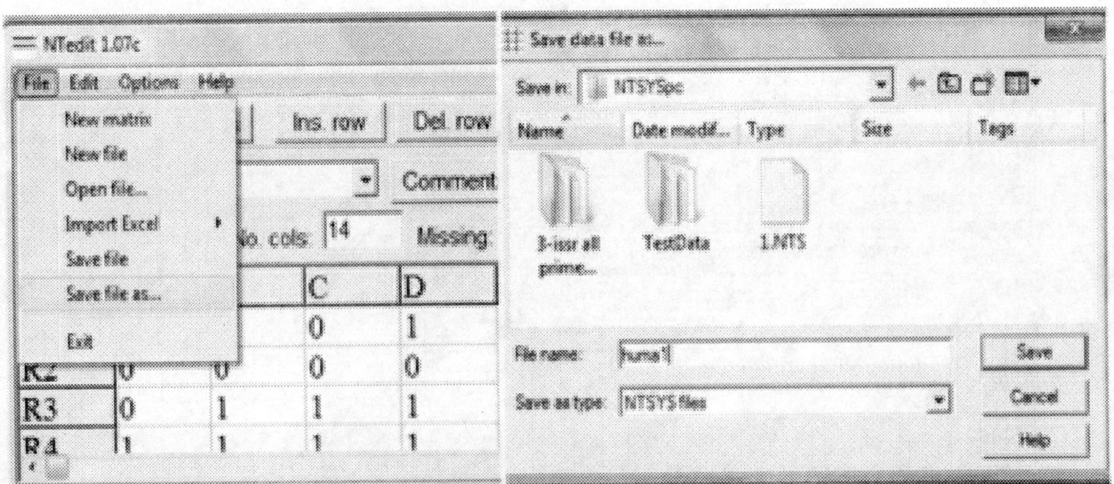

Fig. S4. Saving files in NTSYS folder.

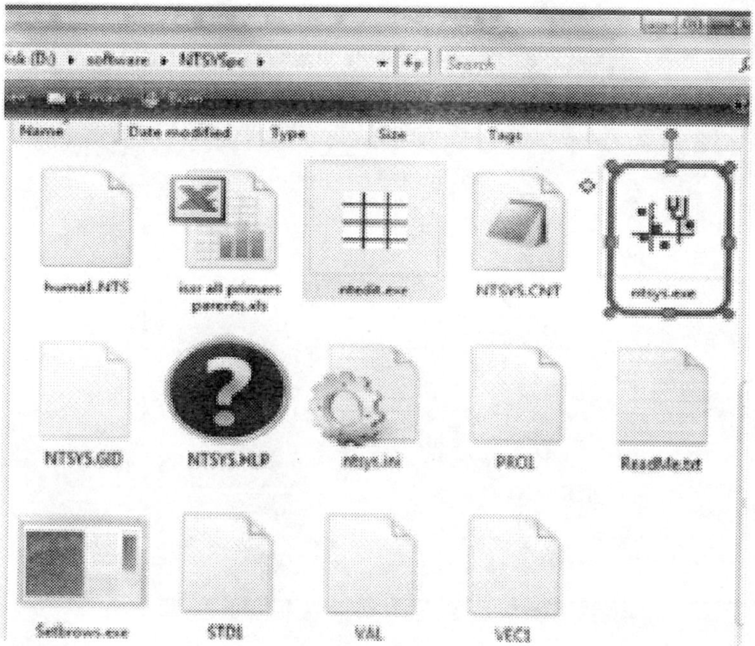

Top

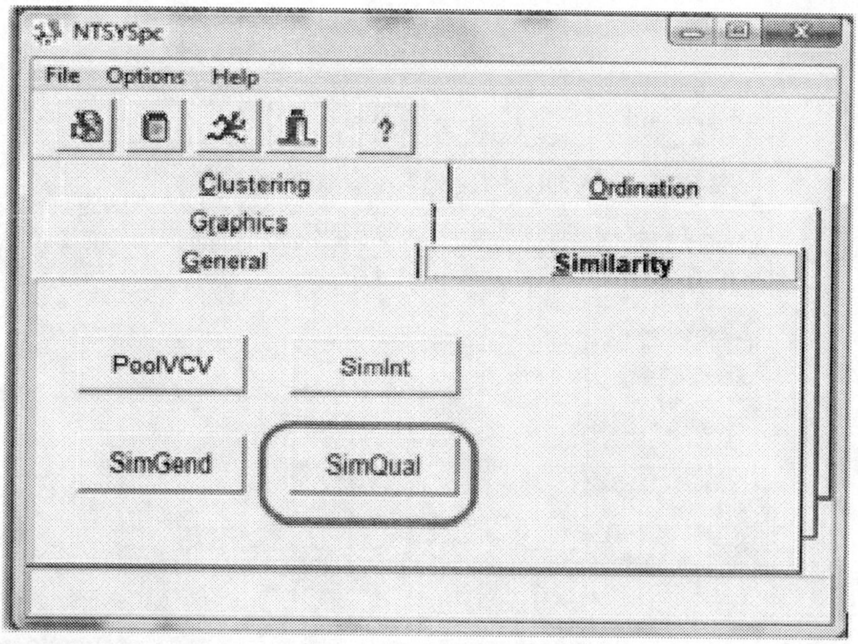

Bottom

Fig. S5. Open NTSYS folder and click on the button marked as RED.

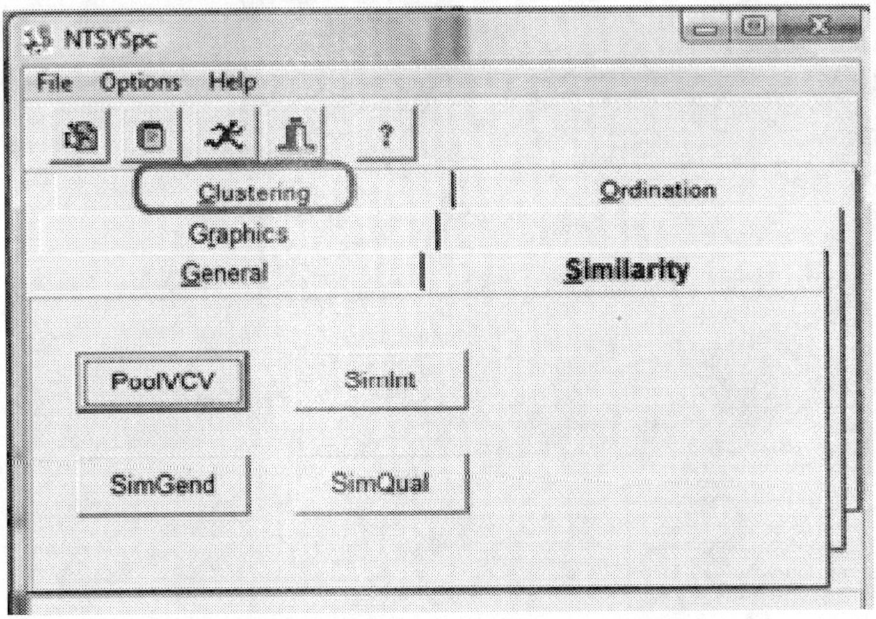

Fig. S6. Use SimQual to compute distance coeffcients.

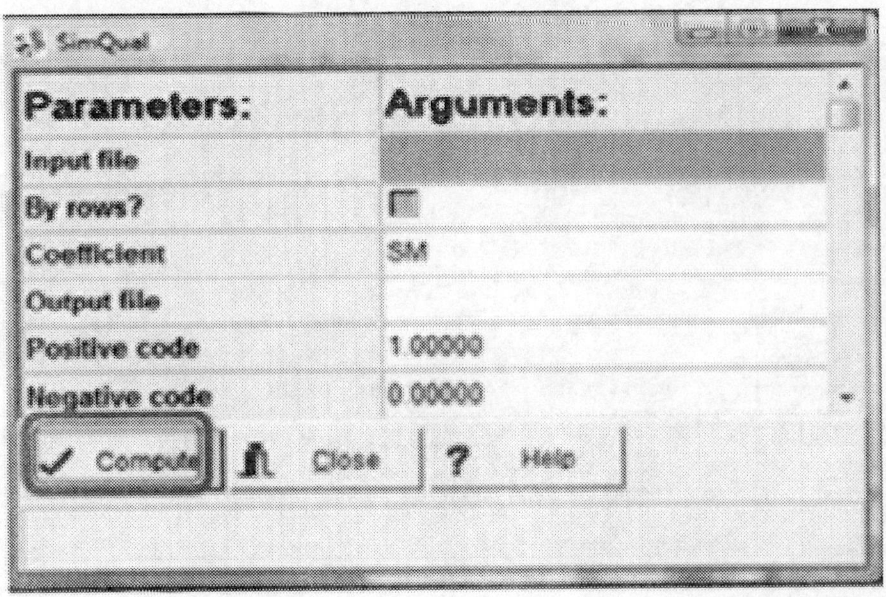

Fig. S7. Report file of the analysis.

Step 7. Click on Clustering button, a window will appear. Now click on SAHN button as shown in Fig. S9. SAHN refers to Sequential, Agglomerative, Hierarchical, and Nested clustering methods, as defined by Sneath and Sokal (1973). Fill the parameter in opened window which appears after clicking on the SAHN (Fig. S10). Here the input file is the output file of Similarity matrix. Give the name to output tree file. Change the setting of Clustering method, in case of ties as find or warn. Then, click to compute button as shown. A window will appear giving the report file. Now close this window.

Step 8. Open the Main Window. Click the Graphics button and then the Tree plot button (Fig. S11). Give the output file of clustering analysis i.e input file of tree plot will be the output file of clustering. Then, click on compute. A tree (dendrogram) will appear in the window (Fig. S12). Save the file.

Go to option menu and change the settings as required i.e. changing font size, color etc. and then save your file.

Step 9. Getting Similarity matrix.

Click the general button. Many options will appear. Click output button. Type the name of input file. i.e. your similarity file (Fig. S13). Change the settings like page width, decimal, and no. of pages etc. according to your choice.

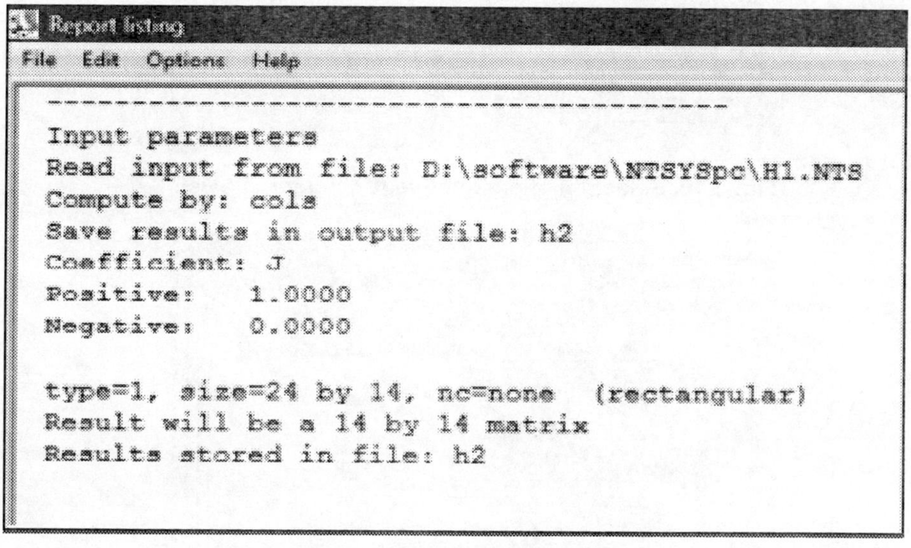

Fig. S8. Report file of the analysis.

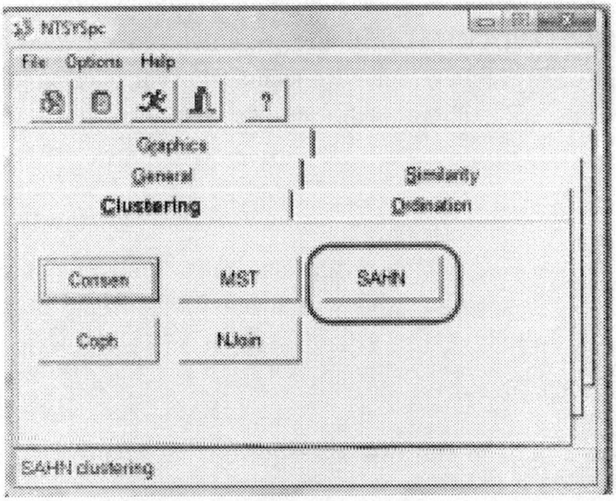

Fig. S9. Clustering options.

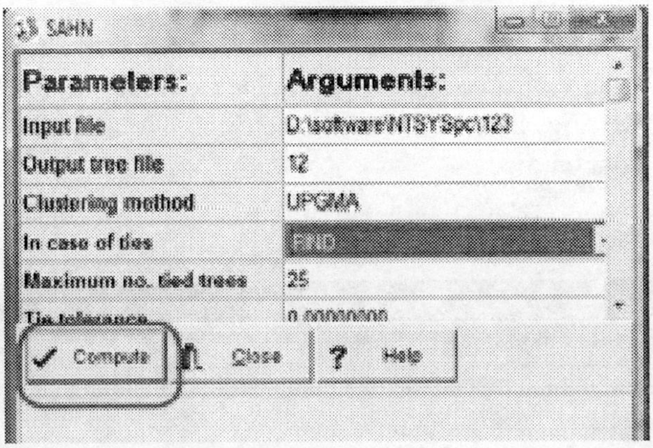

Fig. S10. Selection of parameters for cluster analysis.

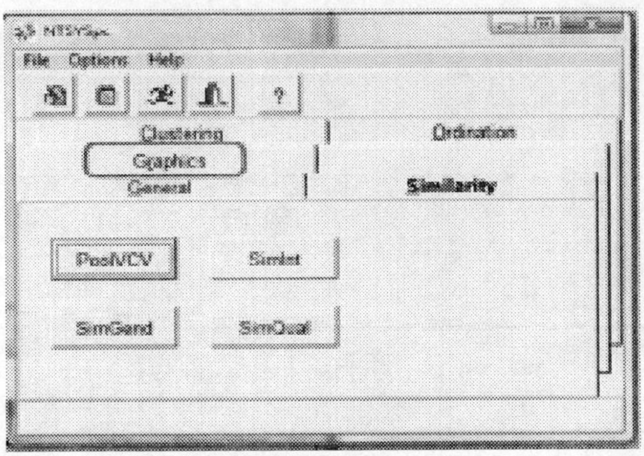

Fig. S11. Computing phylogenetic tree.

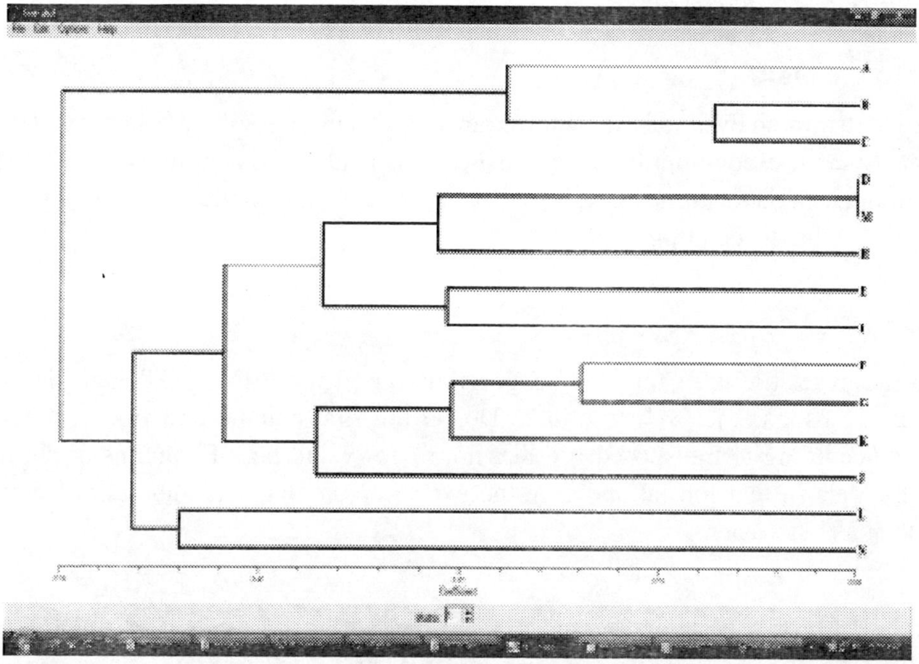

Fig. S12. Output of the data in the form of a tree.

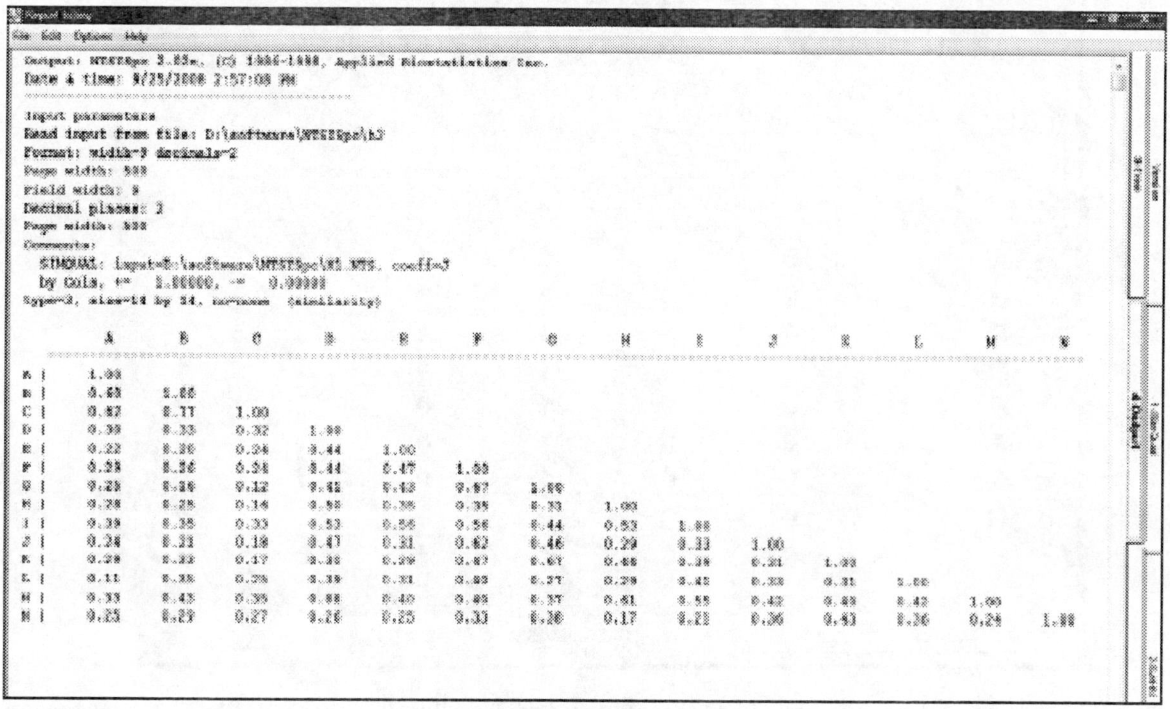

	A	B	C	D	E	F	G	H	I	J	K	L	M	N
A	1.00													
B	0.63	1.00												
C	0.87	0.77	1.00											
D	0.30	0.33	0.32	1.00										
E	0.22	0.30	0.24	0.44	1.00									
F	0.23	0.36	0.24	0.44	0.47	1.00								
G	0.23	0.16	0.12	0.42	0.43	0.67	1.00							
H	0.28	0.28	0.14	0.50	0.30	0.35	0.33	1.00						
I	0.30	0.35	0.33	0.53	0.56	0.56	0.44	0.53	1.00					
J	0.28	0.31	0.18	0.47	0.31	0.42	0.46	0.29	0.33	1.00				
K	0.28	0.33	0.17	0.38	0.29	0.67	0.67	0.48	0.38	0.31	1.00			
L	0.11	0.38	0.26	0.39	0.31	0.83	0.27	0.29	0.43	0.33	0.31	1.00		
M	0.33	0.43	0.35	0.88	0.40	0.48	0.37	0.41	0.55	0.42	0.43	0.42	1.00	
N	0.25	0.29	0.27	0.26	0.25	0.33	0.30	0.17	0.21	0.36	0.43	0.36	0.24	1.00

Fig. S13. Similarity matrix of the data.

B. Bootstrap analysis

To perform Bootstrap analysis of the dendrogram (tree), another software tool available free of charge to the scientific community. This can be downloaded from http://www.irri.org/science/software/winboot.asp. Bootstrap analysis of binary data helps in determining the confidence limits of UPGMA-based dendrograms.

Data input file

The data in the excel file used for NTSYS-pc is also used for WINBOOT analysis with minor modifications. Transpose the excel file data. Delete the first column and first row. Change the first row data according to transposed file like no. of rows and no. of columns as shown in Fig. S 14. Save the data in text format and give the extension as .dat. The Tab delimited format has been used to make it easier.

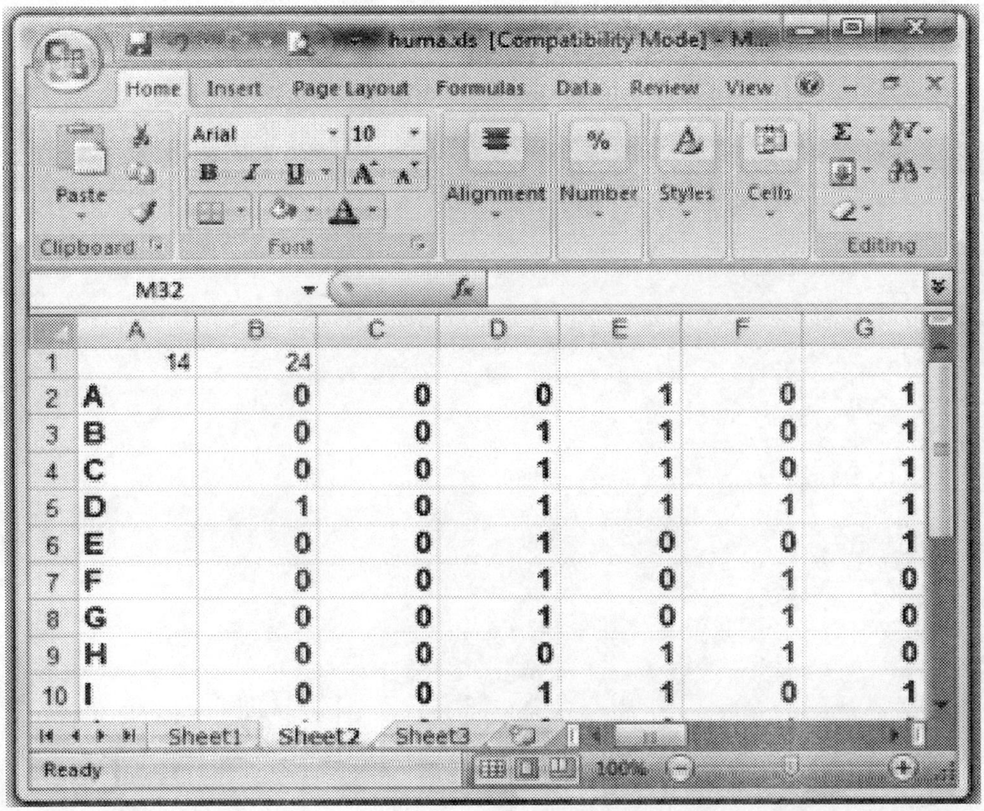

	A	B	C	D	E	F	G
1	14	24					
2	A	0	0	0	1	0	1
3	B	0	0	1	1	0	1
4	C	0	0	1	1	0	1
5	D	1	0	1	1	1	1
6	E	0	0	1	0	0	1
7	F	0	0	1	0	1	0
8	G	0	0	1	0	1	0
9	H	0	0	0	1	1	0
10	I	0	0	1	1	0	1

Fig. S14. Data input file for bootstrap analysis using winboot software.

Step10. Double click on Winboot icon in the program manager. The Winboot window will appear on the screen. To specify **Input file name**, type name of the file directly in the input file name box or by clicking on the browse button (Fig. S15).

Step11. Select types of coefficient and bootstrapping replicates from the drop down box. For instance, Dice coefficient and 1000 bootstrap samples are selected in Fig. S16. Click on compute button to start the bootstrapping process.

The **Coefficient** choice depends on the nature of marker used and the purpose of study. For dominant markers, Jaccard and for co-dominant markers, Dice coefficient can be used. Edit box **Samples** denote the number of iteration to be carried out. The value of the bootstrap and the number of bootstrap samples or replications will determine the accuracy of bootstrap analysis. At 400 replications, the bootstrap estimates would be 1 % where as, 99% accuracy can be achieved at 2000 replications. For random sampling of the matrix the '**Random number seed**' option is used. By default this option should be off and should be activated when program terminates immaturely on some illegal operation, like dividing by zero. Keep changing the seed value until the one that works.

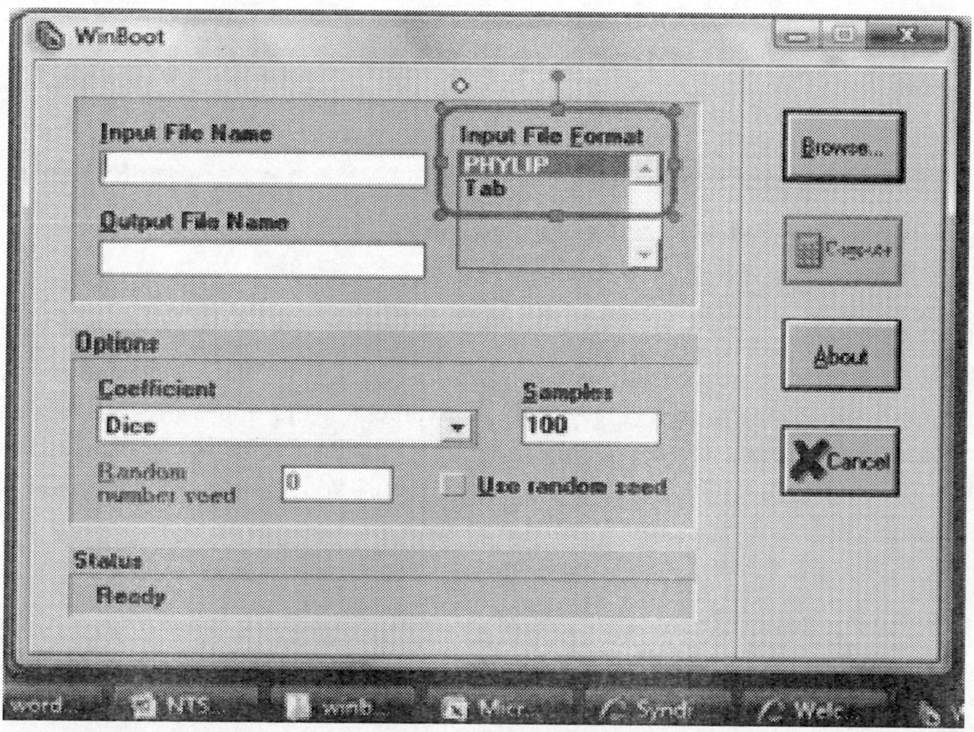

Fig. S15. Winboot window for selecting file format and to browse the file .

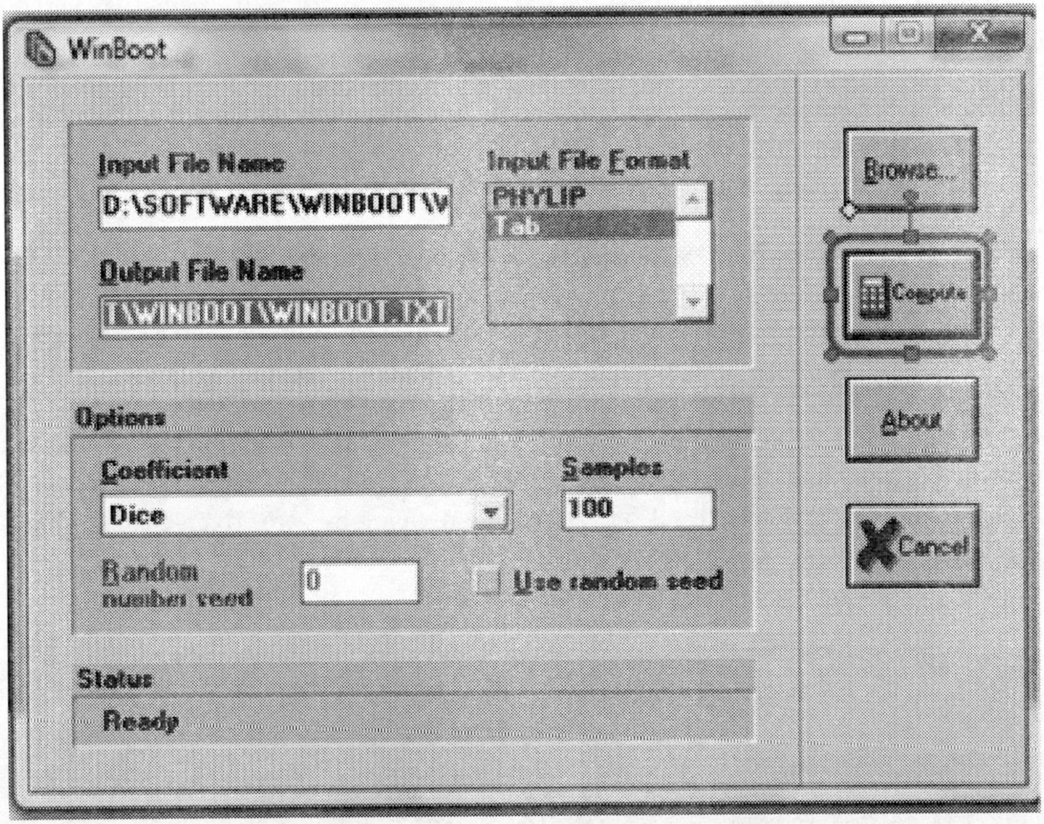

Fig. S16. Selection of specific coefficient and boot strap replicates (samples).

Step 12. Once analysis is over close this window. An output text file is saved in the Winboot folder containing phylogenetic tree (Fig. S17). The values on the nodes of each branch is the bootstrap values which shows the confidence level of each cluster and its statistical significance. More than 75% bootstrap value means the cluster is robust and statistically significant.

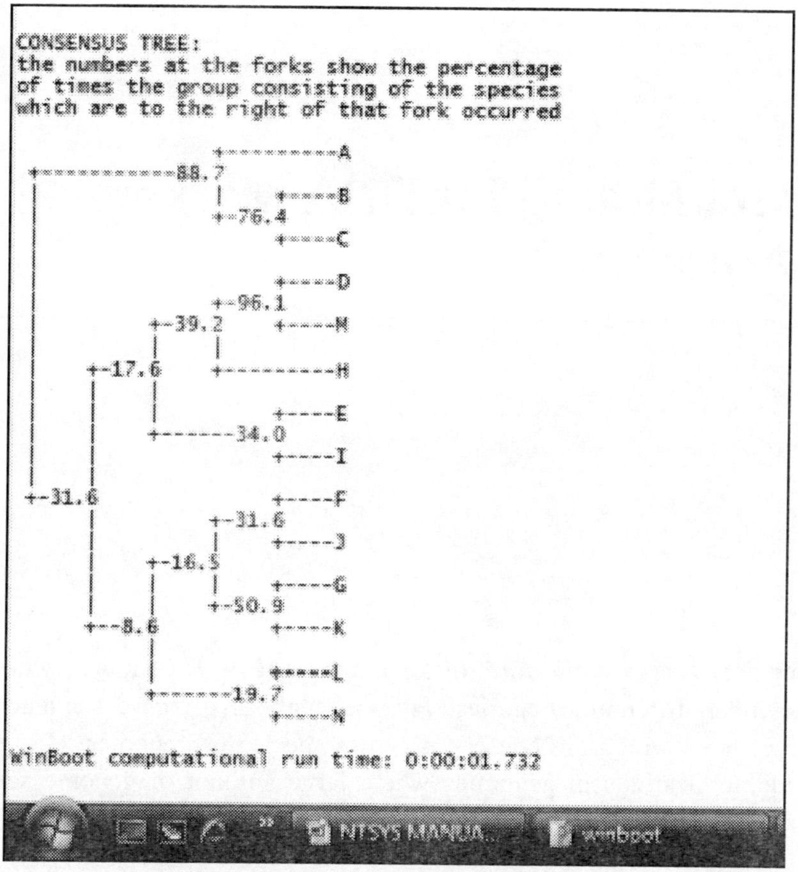

Fig. S17. Output file of the winboot analysis.

Suggested Readings

Rohlf F.J. (1993), NTSYS-PC: Numerical taxonomy and multivariate analysis system. Version 1.80. Exeter Software New York.

Tanksley S.D., N.D. Young, A.H. Paterson, M.W. Bonierbale (1989), RFLP mapping in plant breeding: New tools for an old science. *BioTechnology*, 7: 257-261.

Vos P., R. Hogers, M. Bleeker, M. Reijaus, T. Van de Lee, H. Hornes, A. Frijters, J. Pot, J. Peleman, M. Kuiper, and M. Zabau (1995), AFLP: A new technique for DNA fingerprinting. *Nucl. Acid Res.*, **21**: 4407-4414.

Williams J.G.K., A.R. Kubelik, K.J. Livak, J.A. Rafalski, & S. V. Tingey (1990), DNA polymorphisms amplified by arbitrary primers are useful as genetic markers. *Nucl. Acids Res.*, **18**: 6531-6535

Yap I. and R.J. Nilson (1996), Winboot : A program for performing bootstrap analysis of binary data to determine the confidence limits of UPGMA based dendrograms. IRRI Discussion Paper Ser. 14. Manila, Phillipines.

11

Data Mining for DNA Markers Discovery

Genome sequence data is also in the form of long strings of A, T, G and C which are arranged in the databases under different categories. For the analysis of these sequences, we always look for specific patterns which are conserved across species or genomes. Hence, data mining approach has wide applications in genomics where large amount of genome sequence data is available in the public domain.

WHAT IS DATA MINING?

Data mining is the exploration of large data sets to discover similar patterns and statistically significant events and structures. This approach is based on standard data mining algorithms which include clustering algorithms, tree-based classifiers and association rules. In data mining one has to perform different tasks like prediction of an item class, clustering of data and finding association among different clusters. These clusters are then used in the facilitation of knowledge discovery by describing specific groups. These groups are then further analyzed to find any variation between them and finally establishing statistically significant relationships.

Rapid advancement and development in genome research in the recent past has resulted in the generation of large data set of DNA and protein sequences from different prokaryotes and eukaryotes. Data mining would be one of the important approach of making sense of the DNA sequences of many organisms available in public domain in the post-genomic era. In case of sequence resources of plants, bioinformatics tools can be used for data mining to discover many useful genes for agriculturally important traits. These identified genes can be further characterized at molecular level and used in validation experiments by using functional genomics approaches and then incorporated in crop improvement programs. Based on gene information from the databases, genome wide expression analysis of the genes can be performed to understand their interaction at molecular level. The data mining approach would also be very helpful in developing single nucleotide polymorphisms (SNPs) and simple sequence repeat (SSR) markers. SNPs have been reported to be the most frequent form of DNA variation and are considered next generation of genetic markers that can be used in precision breeding. Various strategies like experimental and *in silico,* can be employed for the discovery and mapping of SNPs. However, experimental approaches used for SNP detection involves large number of laborious, complex and expensive steps. Hence, *in silico* discovery of SNPs has been proposed as an alternative strategy which can make use of various computer softwares and large data set obtained from different genome sequencing projects. Various SNP detection tools and parameters can be optimized to achieve specific goals.

ESTs are considered an important genomic reserve for mining DNA markers based on microsatellites or simple sequence repeats (SSRs). The SSRs or microsatellites, are present and distributed in the genomes of all eukaryotes. Due to the abundance and specificity of SSRs, these are important DNA markers for genetic mapping and population studies. The SSRs are tandem repeats of mono-, di-, tri-, tetra- and penta-nucleotides with different lengths of repeat motifs. High levels of genetic variation are obtained because of the differences in the number of tandemly repeating units of a locus. The important features of SSR markers coupled with their ease of detection have made them useful molecular marker in different crops. Therefore, detection of SSRs in the unigenes and ESTs of different plant species may help in designing new set of DNA markers and may provide more insight in the evolution of these species.

DNA markers are landmarks spread throughout the genome and used in constructing genetic linkage maps in plants, construction of physical maps, map based cloning of the genes and orientation of BAC/PAC clones in the large genome sequencing projects. The term marker is used very broadly to describe any observable variation that results from an alteration, or mutation, at a single genetic locus. A marker may be used as one landmark on a map if, in most cases, that stretch of DNA is inherited from parent to offspring in a typical Mendalian fashion. Markers are commonly found within the non-coding regions of genes and are used to detect unique regions on a chromosome. However, when the markers are found within the genes, these are called functional or gene-based markers.

These DNA markers can be shared with the plant breeders for use in marker assisted selection (MAS) for precision breeding and for gene pyramiding to develop cultivars with durable and multiple resistance to various stresses.

DATA MINING FOR DNA MARKERS

Markers can be designed *in-silico* either from the genomic sequences, ESTs or gene sequences of plants. The process of *in-silico* identification of DNA markers can be explained as a chronology of the processes as shown in Fig. 1.

DATA MINING FOR SSR MARKERS

Identification of SSRs is done using SSR Identification Tool (www.gramene.org). It is a perl program used to find out the desired length of repeats in the given sequences (Fig. 2).

Once repeat region is identified, primers are designed flanking to the repeats. Primer 3 software is commonly used to design custom primer pairs from the region flanking to the repeat elements. Primer3 picks primers for PCR reactions, considering different criteria: the oligonucleotide melting temperature, size, GC content and primer-dimer possibilities, PCR product size, positional constraints within the source sequence and other miscellaneous constraints. All of these criteria are user-specifiable as constraints and some are specifiable as terms in an objective function that characterizes an optimal primer pair. Whitehead Institute for Biomedical Research provides a web-based front end to Primer3 at http://fokker.wi.mit.edu/cgi-bin/primer3/primer3_www.cgi

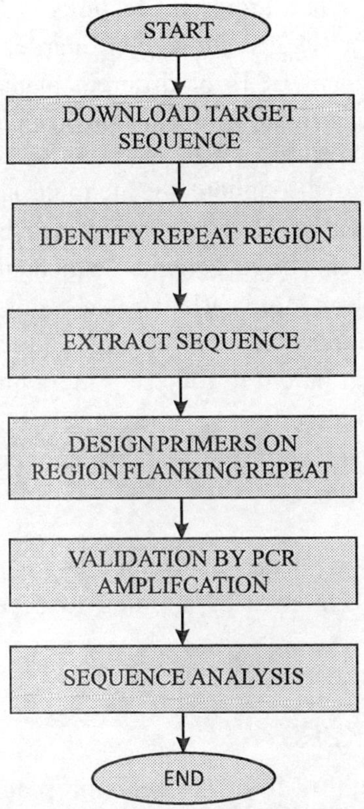

Fig. 1. Process flow diagram of *in-silico* discovery of DNA markers.

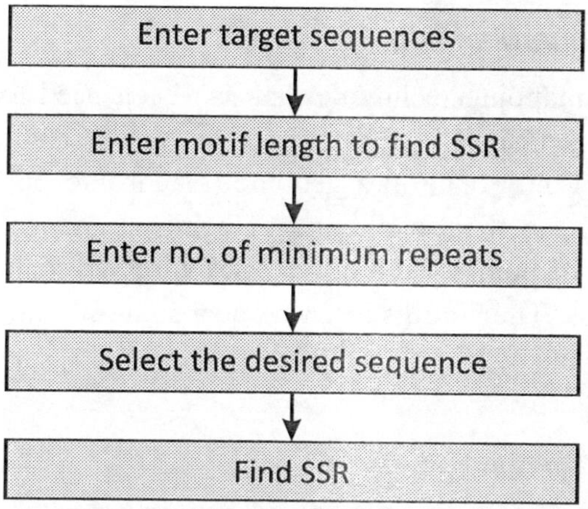

Fig. 2. Process flow diagram showing SSR detection process using SSRIT software.

Identification of STS markers can be performed by using electronic PCR (e-PCR). E-PCR is a computational tool that is used to identify sequence tagged sites (STSs), within the DNA sequences. e-PCR looks for potential STSs in DNA sequences by searching for subsequences that closely match the PCR primers and have the correct order, orientation and spacing that could represent the PCR primers used to generate known STSs. The highly specific and sensitive PCR provides the basis for STSs, unique landmarks that have been used widely in the construction of genetic and physical maps of different genomes. The significance of this technique can be seen by considering that it is possible to determine the map location of a new sequence without performing a single experiment in the laboratory.

For high throughput SSR repeat identification in the genomic sequences an important tool known as MISA - Microsatellite Identification Tool is used. Its availability and use has been explained in the exercise given in this chapter.

Use of EST for DNA Markers Development

ESTs are small pieces of DNA sequence (usually 200 to 500 nucleotides long) that are generated by sequencing either one or both ends of an expressed gene. The idea is to sequence bits of DNA that represent gene signatures in certain cells, tissues, or organs from different organisms and use these "tags" to fish a gene out of a portion of chromosomal DNA by matching base pairs. The challenge associated with identifying genes from genomic sequences varies among organisms and is dependent upon genome size as well as the presence or absence of introns. ESTs are first assembled into contigs to remove redundancy and then SSR markers are identified within these contigs using computational tools.

Use of ESTs as a tool for Gene Mapping

Currently, the most powerful mapping technique that has been used to generate many genome maps, relies on Sequence Tagged Site (STS) mapping. An STS is a short DNA sequence that is easily recognizable and occurs only once in a genome (or chromosome). The 3' ESTs serve as a common source of STSs due to their likelihood of being unique to a particular species and provide the additional feature of pointing directly to an expressed gene. Similarly, ESTs can be clustered in the form of contigs. The contigs are used as a source sequence for the identification of SSR markers as described in Fig. 2.

ESTs as Gene Discovery Resource

The ESTs represents a signature of a gene, which is expressed. EST resources have been a powerful tool in the hunt for genes involved in many different traits. ESTs also have a number of practical advantages such as, their sequences can be generated rapidly and inexpensively, only one sequencing experiment is needed for each cDNA generated and they do not have to be checked for sequencing errors since mistakes do not prevent identification of the gene from which the ESTs are derived.

Data mining for SNP markers

A Single Nucleotide Polymorphism, or SNP (pronounced "snip"), is a small genetic change, or variation, that can occur within an organism's DNA sequence. The genetic code is specified by the four-nucleotide "letters" A (adenine), C (cytosine), T (thymine), and G (guanine). SNP variation occurs when a single nucleotide, such as an A, replaces one of the other three-nucleotide letters—C, G, or T.

An example of a SNP is the alteration of the DNA segment AAGGTTA to ATGGTTA, where the second "A" in the first snippet is replaced with a "T". For example, SNPs occur in the rice genome at a frequency of 1 in every 450 bp. Since only about 3 to 5 percent of an organism's DNA sequence codes are responsible for the production of proteins, most SNPs are found outside of "coding sequences". SNPs found within a coding sequence are of particular interest to researchers since they are more likely to alter the biological function of a protein. With the recent advances in technology, coupled with the unique ability of these genetic variations to facilitate gene identification, there has been a recent flurry of SNP discovery and detection.

In silico SNP Detection Tools

Identification of SNPs and mutations is important for the identification of trait they influence. PCR resequencing is the method of choice for *de novo* SNP discovery. However, manual curation of putative SNPs has been a major bottleneck in the application of this method to high-throughput screening. Therefore, it is critical to develop a more sensitive and accurate computational method for automated SNP detection. A software tool, SNPdetector is used, for automated identification of SNPs and mutations in fluorescence-based resequencing reads. SNPdetector was designed to model the process of human visual inspection and has a very low false positive and false negative rate. The analysis of large and diverse test data sets demonstrated that SNPdetector is an effective tool for genome-scale research and for large-sample clinical studies. SNPdetector runs on Unix/Linux platform and is available publicly (http://lpg.nci.nih.gov).

SNP identification software is known as PolyPhred. PolyPhred is a program that helps to accurately identify heterozygous sites in the sequences produced by sequencing PCR products with fluorescence-based chemistries such as dye labeled terminators or dye-labeled primers. The program compares sequence traces and searches for homozygotes and heterozygotes. It compares fluorescence-based sequences across traces obtained from different individuals to identify heterozygous sites for single nucleotide substitutions (Fig. 3).

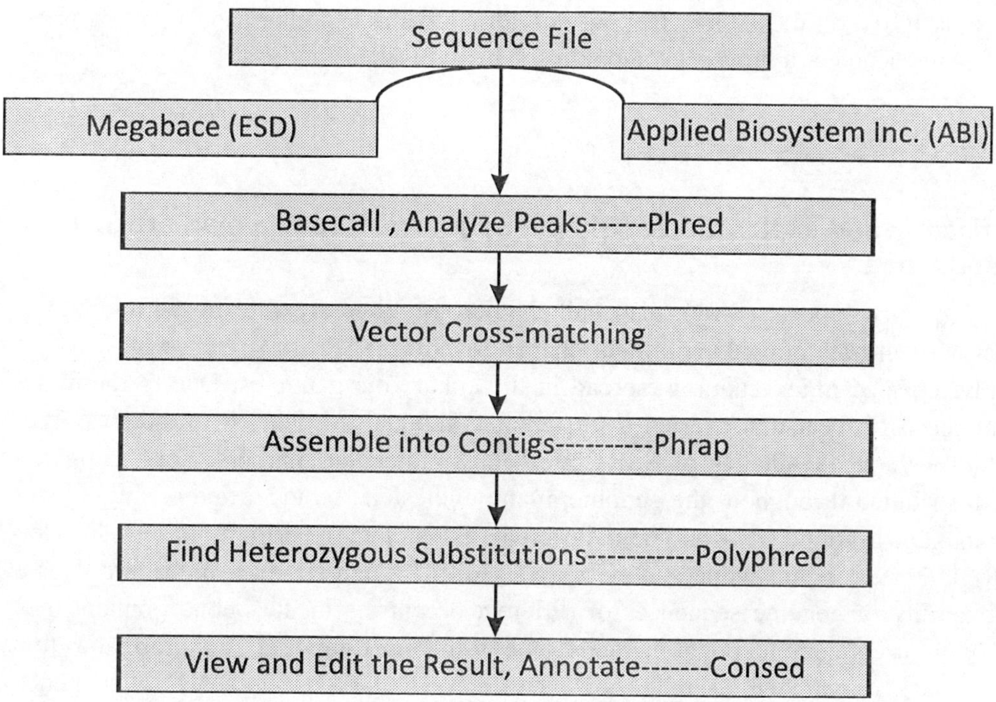

Fig. 3. A flow diagram of SNP detection process by PolyPhred.

Basics of SNPs Detection using PolyPhred

1) A significant drop in fluorescence peak height at a variant site when sequence traces obtained from homozygous individuals are compared to traces from heterozygous individuals (theoretical drop is expected to be 50%).
2) The presence of a second fluorescence peak in sequence traces from heterozygous individuals. PolyPhred scans for these two features when sequence traces are being compared to detect heterozygotes among homozygotes.

The following rank determining factors should be taken into consideration while identifying SNPs:

1) The ratio of the areas under the two peaks (called the area ratio).
2) The ratio of the actual height of one of the peaks to the height of a hypothetical homozygous peak (called the normalization ratio). The peak that is used corresponds to the consensus base at the position.
3) The average quality, assigned by Phred, of the sites flanking the heterozygous site (the two sites immediately adjacent to the heterozygous site are excluded from the average, as Phred typically reduces their quality due to the heterozygous site itself). After assigning an initial rank based on these three factors, PolyPhred examines other aspects of the trace, such as the presence of a third peak and adjusts the rank accordingly.

EXERCISE

Identification and exploration of Simple Sequence Repeats markers in a given DNA sequence.

Simple sequence repeats (SSR), also called microsatellites are the most important molecular markers used in both animals and plant characterization. SSR are stretches of 1 to 6 nucleotide tandemly repeated units randomly spread in the eukaryotic genomes. Due to the high mutation rate, number of repeat units varied in different individuals which led to length polymorphism. There are several advantages of SSRs over other molecular markers. For example, these are evenly distributed throughout the genome, many alleles can be identified at single locus, require little amount of genomic DNA and no radioactivity is required to perform the analysis. Development of SSR markers from eukayotic system is a very difficult and complex procedure. However, with the availability of genome sequences of different organisms in the public domain, these can be identified by using computational methods. One of the most important bioinformatics tools used is known as MISA which is used for the identification of SSRs in the DNA sequences. It has been explained in the following sections.

MISA - Microsatellite Identification Tool

This software is freely available in public domain and can be downloaded from http://pgrc.ipk-gatersleben.de/misa/misa.html site. It allows identification of perfect as well as compound microstallite in the given sequences. After downloading, it can be used on command line as explained in the documentations of the software. Following steps can be used for the searching microsatellites.

Step 1: First download the sequence in which SSRs to be identified as shown below in the NCBI window (Fig. S1). The accession number EE595475 is the sequence of a EST clone of *Cajanus cajan* which is used as an example.

Step 2: The sequence to be analyzed for SSR mining should be in FASTA format and transferred in WordPad file (Fig. S2).

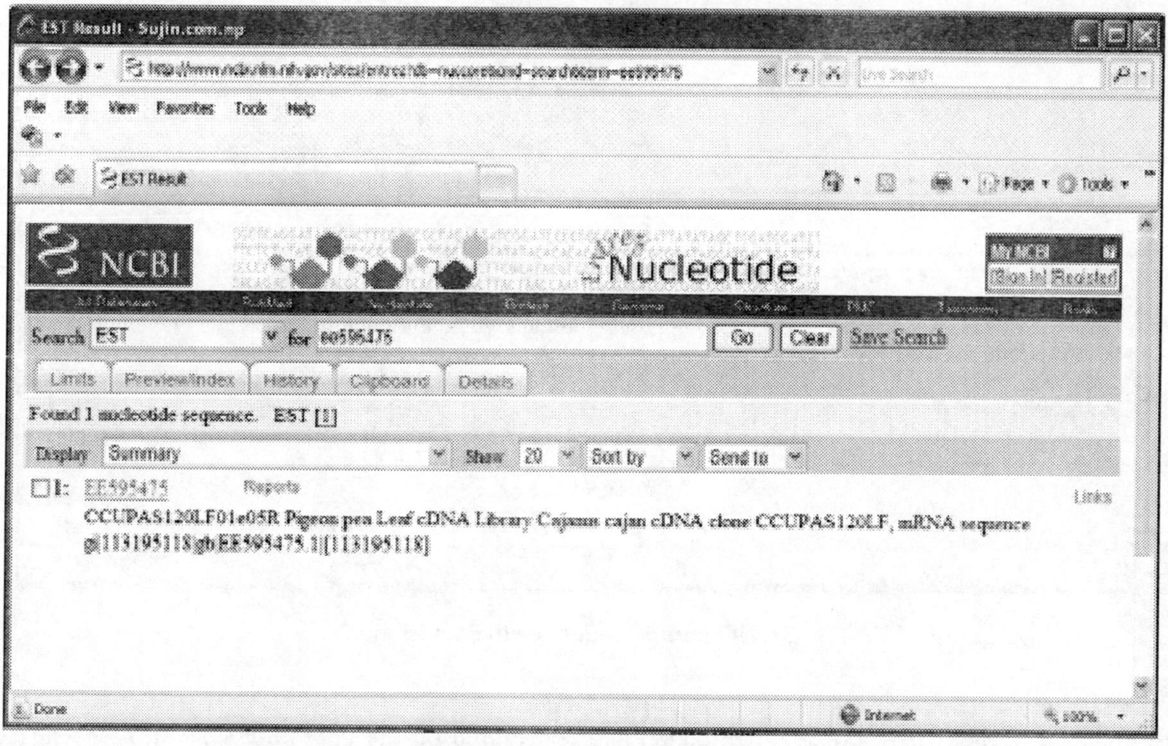

Fig. S1. Download EST sequence from the NCBI database.

Fig. S2. Nucleotide sequence in Fast format.

Step 3: Once the MISA tool is downloaded and configured on the operating system, it is ready for use. The two files misa.ini and misa.pl are generated after configuration to be used for SSR mining. Through command prompt following commands are executed "**perl misa.pl <File name>".** Fig S3 shows the command and the file generated after analysis **File name. Statistics & file name.misa**.

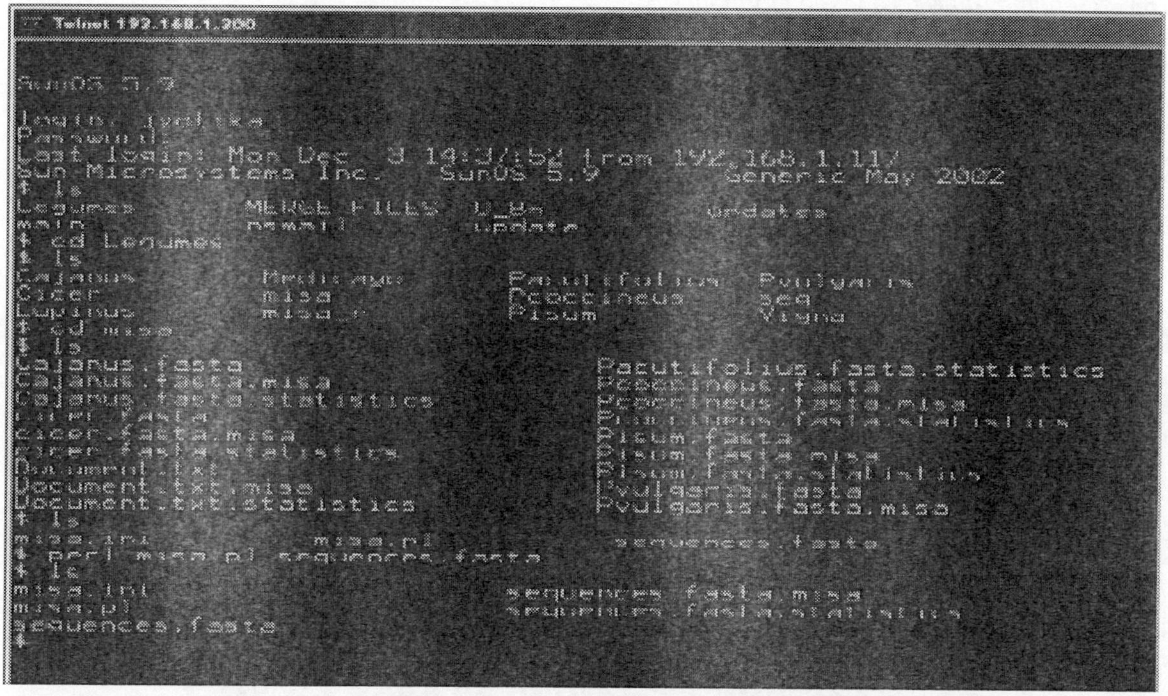

Fig. S3. Different commands and the file generated after analysis.

Step 4: The results generated from MISA tool are automatically stored in two files. The First file is file_name.misa showing the detailed SSR analysis comprising monomers, dimers, trimers etc.(Fig. S4a). In this output, file following information are stored about the analyzed sequence.

ID: The main header of the sequence used for SSR mining.

SSR nr: Number of SSRs found in the given sequence.

SSR type: The type of SSR mined i.e., whether it's a monomer, dimer, trimer, tetramer, pentamer or hexamer.

SSR: The motif type or the repeat sequence that comprises the sequence of SSR along with number of repeats. In this example $(CGC)_5$ denotes (CGC) as the repeat sequence and the number 5 denotes the number of times CGC is repeated within the SSR.

Size: The total length of the SSR motif, which is equal to "Number of bases in the motif X Digit number with the motif". Here the motif found was $(CGC)_5$, so the length of the size is equal to 3×5 =15.

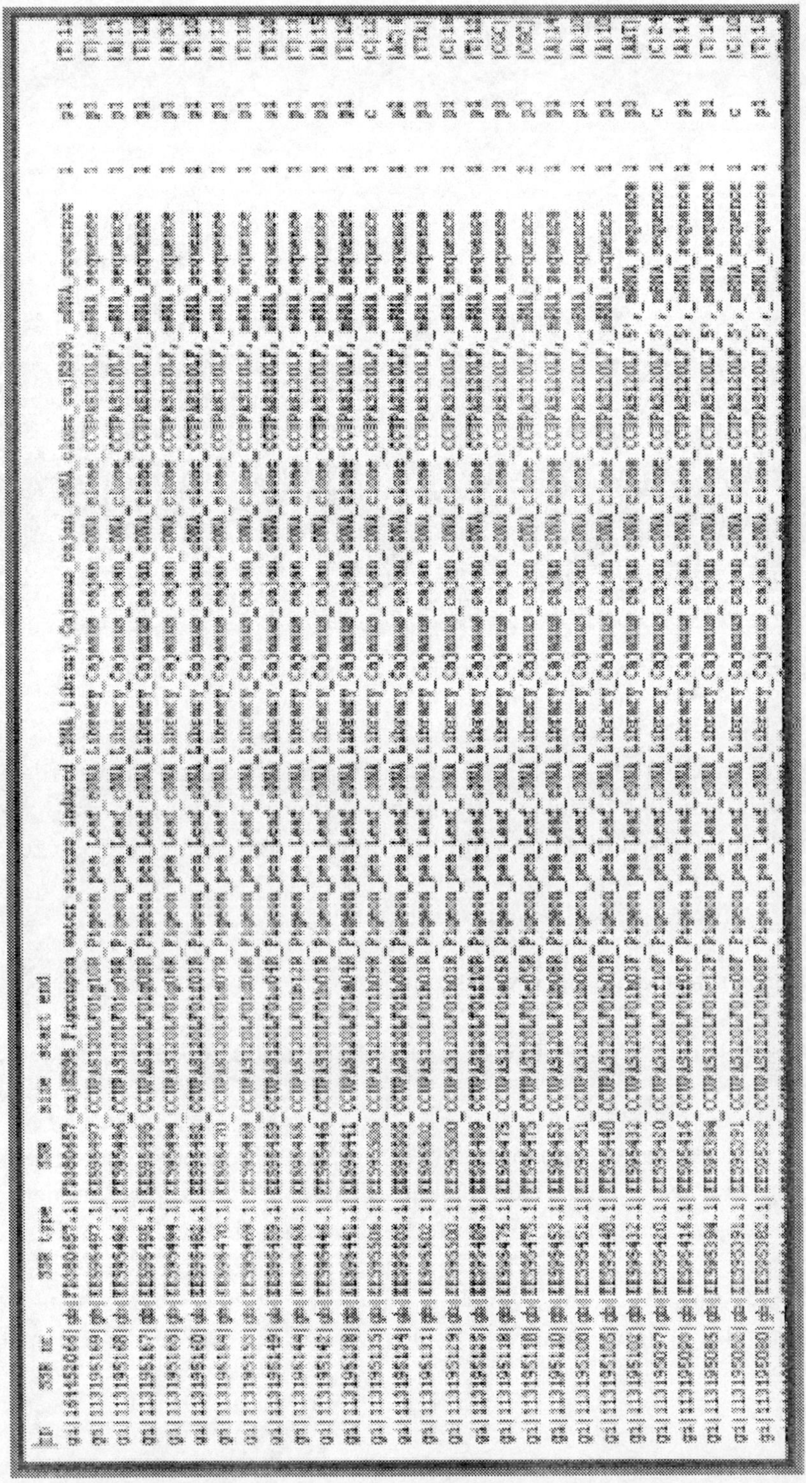

Fig. S4a. Output window of the "file_name.misa" obtained after analysis.

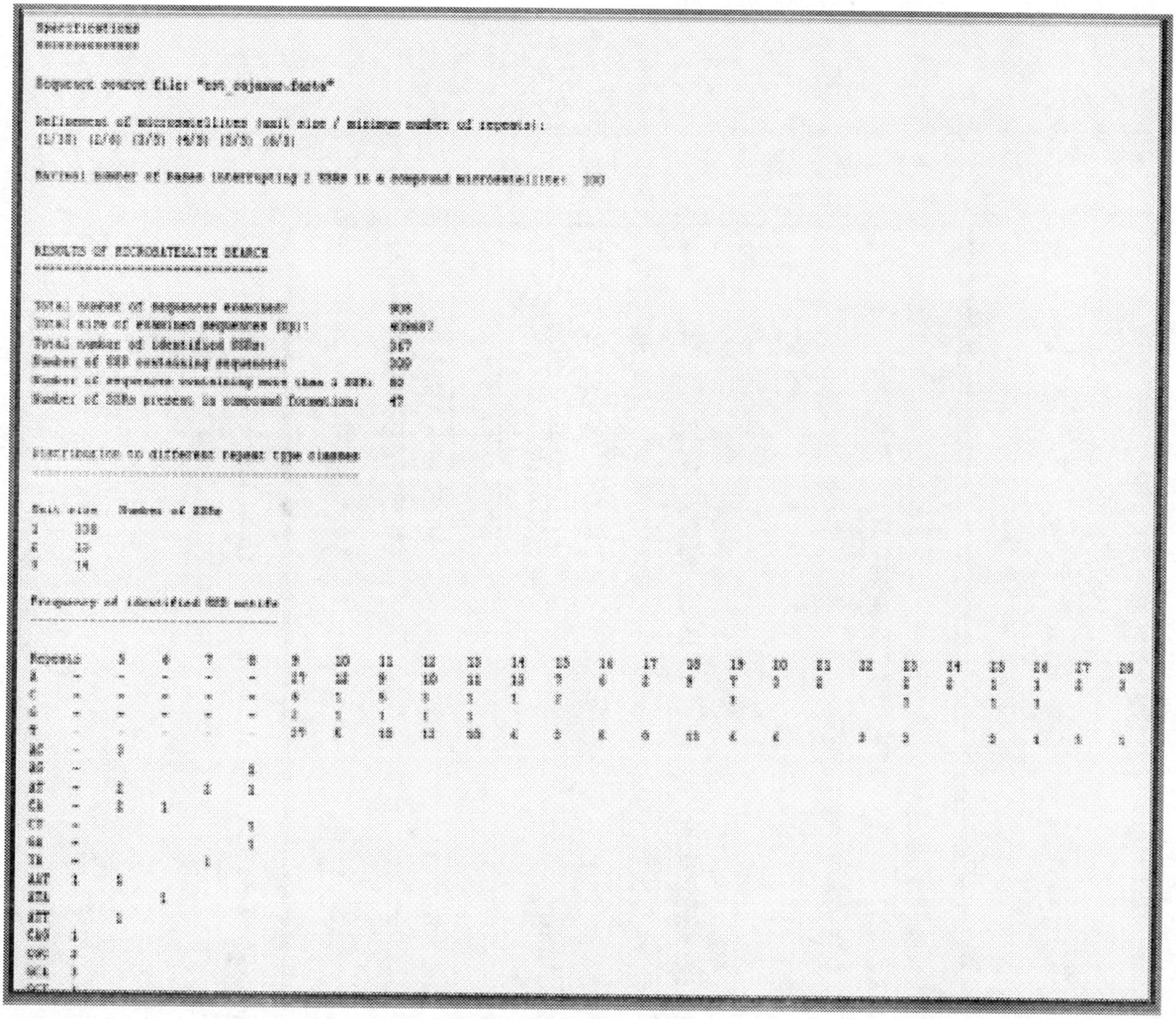

Fig. S4b. Results of the SSR mining.

Start: The position where the SSR begins in the input sequence.

End: The position where the SSR ends in the input sequence.

The second file is "file_name.statistics" showing the statistical analysis of SSR mining results (Fig. S4b).

Step 5: After identification of SSR motifs within a sequence, primers flanking to the SSR motifs are designed. For this analysis, we added 100 nucleotides to right and left sides of the flanking region of SSR motifs using excel software (Fig. S5).

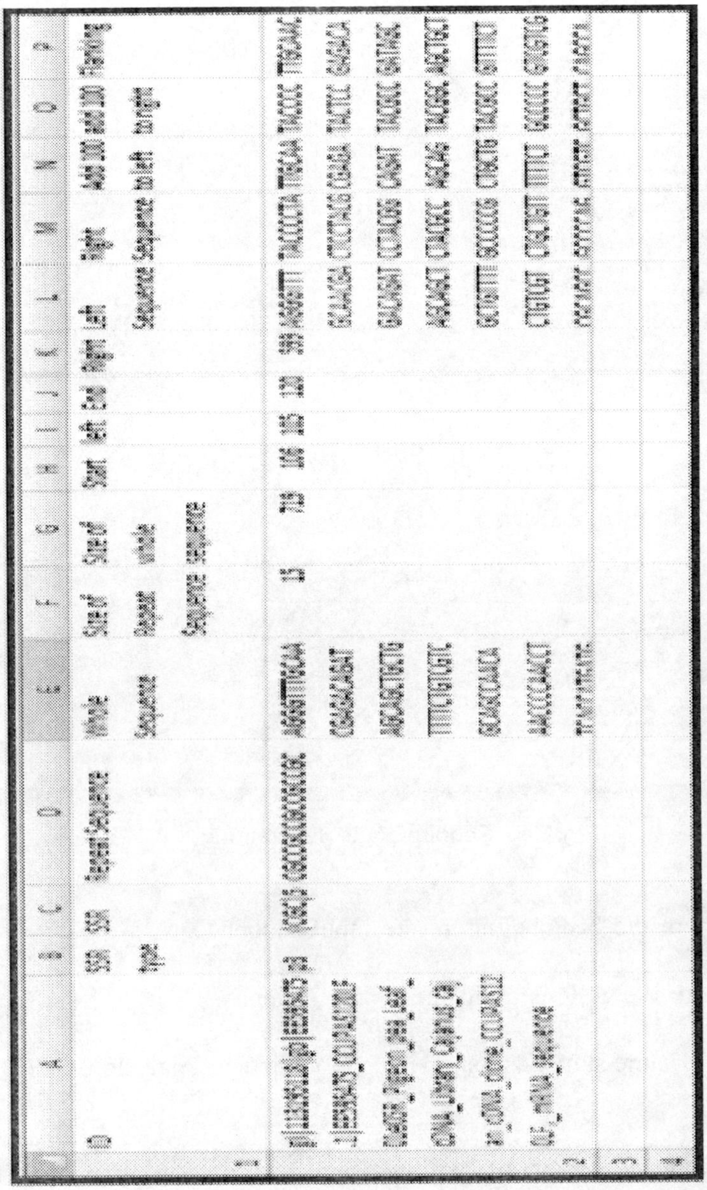

Fig. S5. Output window showing SSR motifs along with additional nucleotides.

Step 6: After adding the flanking sequences to the SSR, the sequence is then used for primer designing using Primer3 software, which is available on http://frodo.wi.mit.edu/ as shown in figure S6.

Step 8: The results generated from Primer3 Software and their Tm (Melting Temperature), GC content, product size and sequences of left and right primers can be obtained. (Fig. S7). Besides, this software also give additional primer options which can also be tried for PCR amplification if the first option fails.

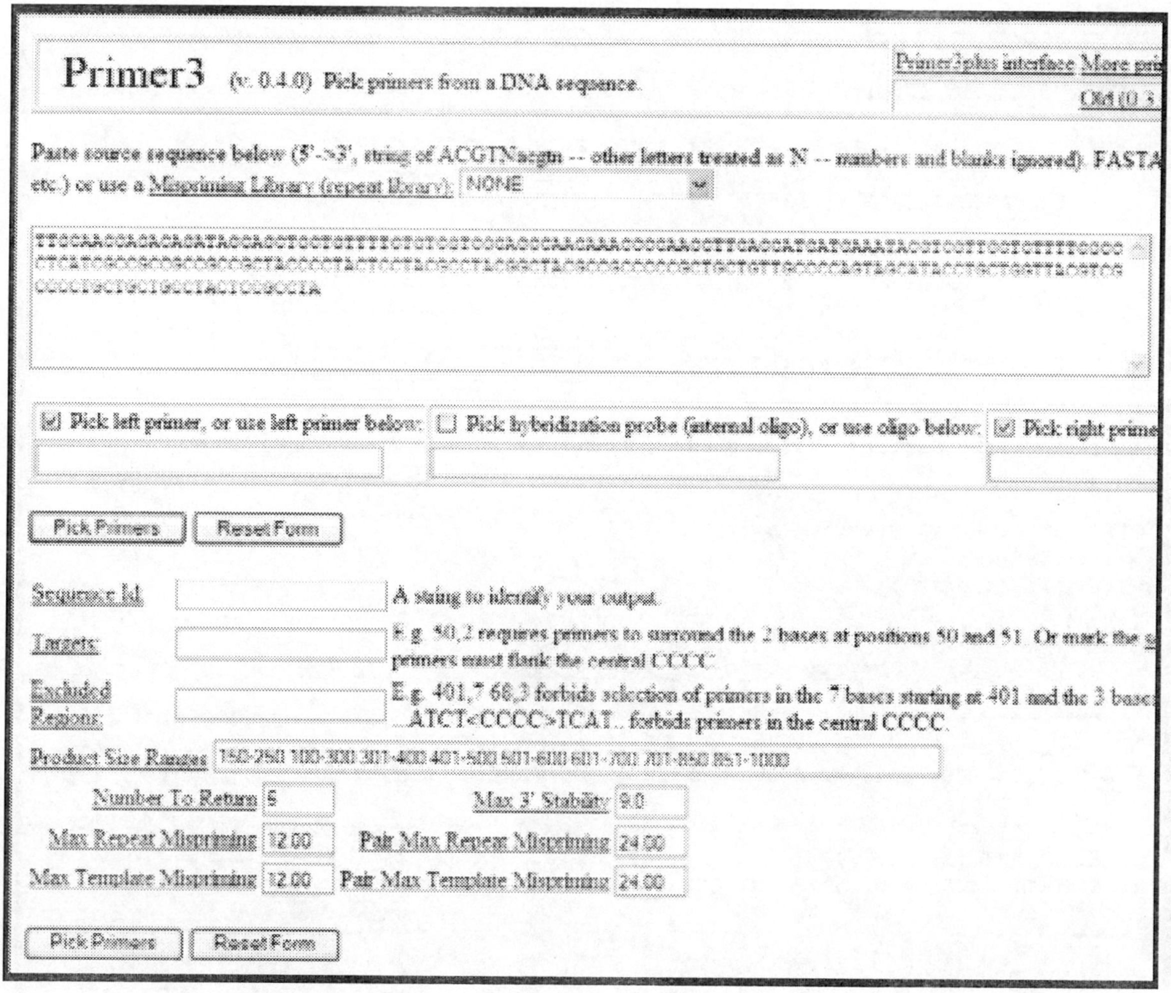

Fig. S6. Primer 3 output window. Sequence is pasted in the window and primer designing criteria is selected before asking program to pick primer.

Primer3 Output

```
No mispriming library specified
Using 1-based sequence positions
OLIGO            start  len     tm     gc%   any    3'  seq
LEFT PRIMER         19   20  59.24   50.00  7.00  0.00 CAGCTGCTGTTTTCTGTCGT
RIGHT PRIMER       176   20  60.91   55.00  4.00  2.00 GGTATGCTACTGGGGCAACA
SEQUENCE SIZE: 215
INCLUDED REGION SIZE: 215

PRODUCT SIZE: 158, PAIR ANY COMPL: 4.00, PAIR 3' COMPL: 2.00          Forward
                                                                      Primer
    1 TTGCAACGAGACAGATAGCAGCTGCTGTTTTCTGTCGTCGCAGCCAACAAACCCCAACCT
                        >>>>>>>>>>>>>>>>>>>>

   61 TCACCATGATGAAATACGTCGTTTGCTCTTTTTGGCCCTCATCGCCGCCGCCGCCGTACCC   SSR
                                                                      markers

  121 CTACTCCTACGCCTACGGCTACGCCGCCCCCCGCTGTGTTGCCCCAGTAGCATACCTGCT
                                          <<<<<<<<<<<<<<<<<<<<

  181 GGTTACGTCGCCCCTGCTGCTGCCTACTCCGCCTA
                                                   Reverse
                                                   Primer

KEYS (in order of precedence):
>>>>>> left primer
<<<<<< right primer

ADDITIONAL OLIGOS
                     start  len     tm     gc%   any    3'  seq

  1 LEFT PRIMER         19   20  59.24   50.00  7.00  0.00 CAGCTGCTGTTTTCTGTCGT
    RIGHT PRIMER       177   20  59.08   55.00  4.00  3.00 AGGTATGCTACTGGGGCAAC
    PRODUCT SIZE: 159, PAIR ANY COMPL: 3.00, PAIR 3' COMPL: 2.00

  2 LEFT PRIMER         19   20  59.24   50.00  7.00  0.00 CAGCTGCTGTTTTCTGTCGT
    RIGHT PRIMER       174   20  61.19   55.00  4.00  3.00 TATGCTACTGGGGCAACAGC
    PRODUCT SIZE: 156, PAIR ANY COMPL: 6.00, PAIR 3' COMPL: 2.00

  3 LEFT PRIMER         20   20  58.24   50.00  5.00  2.00 AGCTGCTGTTTTCTGTCGTC
    RIGHT PRIMER       176   20  60.91   55.00  4.00  2.00 GGTATGCTACTGGGGCAACA
    PRODUCT SIZE: 157, PAIR ANY COMPL: 4.00, PAIR 3' COMPL: 3.00
```

Fig. S7. Output of primer 3 software. The SSR motifs and forward and reverse primer sequences are shown in boxes. The best primer pairs are given at the top.

Suggested Readings

Bhangale, T. R, M. Stephens and D.A. Nickerson (2006), Automatic resequencing based detection of insertion-deletion polymorphisms. *Nat. Genet.* **38**: 1457-62.

Luciano Carlos da Maia, Dario Abel Palmieri, Velci Queiroz de Souza, Mauricio Marini Kopp, Fernando Irajá Félix de Carvalho, and Antonio Costa de Oliveira·(2008), SSR Locator: Tool for Simple Sequence Repeat Discovery Integrated with Primer Design and PCR Simulation. Int. *J. Plant Genomics*. 412696.

Nickerson, D. A., V. O. Tobe and S. L. Taylor (1997), PolyPhred: Automating the detection and genotyping of single nucleotide substitutions using fluorescence-based resequencing. Nucleic Acids Res. **25**: 2745-51.

Rozen, S. and H.J. Skaletsky (2000), Primer 3 on www for general users and for biologist programmers.In: Krawetz S, Misener S (eds) *Bioinformatics Methods and Protocols: Methods in Molecular Biology.* Humana Press, Totowa, NJ, pp 365-386.

Temnykh S, G DeClerck, A, Lukashova, L. Lipovich, S. Cartinhour and S. McCouch (2001), Computational and experimental analysis of microsatellites in rice (Oryza sativa L.): frequency, length variation, transposon associations, and genetic marker potential Genome Res. **11**: 1441-1452.

12

Polymerase Chain Reaction and PCR Primer Design

The polymerase chain reaction (PCR) provides a simple and sensitive method to amplify the specific DNA sequences exponentially by *in vitro* DNA synthesis with the help of flanking primers and thermostable DNA polymerase.

THE PCR TECHNOLOGY

The process of PCR was first invented by Prof. Kary Mullis in 1985 for which he received the Noble Prize in Chemistry for the year 1993. With the advent of PCR, studies on molecular biology and biotechnology have undergone complete transformation. Theoretically, PCR is a very simple three step reaction, which is carried out at three different temperature regimes. The pre–requisites for PCR are,

(i) a template (DNA or cDNA to be amplified),

(ii) a pair of single stranded oligonucleotides known as primers,

(iii) DNA polymerse to catalyze DNA synthesis,

(iv) suitable deoxyribonucleotide triphosphates (dNTPs), PCR buffer and $MgCl_2$ (Table1).

All these components are mixed in a small eppendorf tube and subjected to three steps (cycle) of repeated temperature conditions which are conducted in succession (Table 2). First step is the denaturation of double stranded DNA (template) at a high temperature (Fig. 1). During this step, both the strands of DNA remain separated in the solution until the temperature is lowered. In the second step, temperature is lowered sufficiently to allow primers to anneal on the opposite strands of template DNA flanking the region to be amplified. Since, primers are present in large

Table 1. Concentration of various chemical components used for 50 μl PCR reaction.

Components	Volume (μl)	Final concentration
10X PCR buffer	5	10mM Tris HCl 50mM KCl
10mM dNTP mixture	1	0.2mM each
50 mM $MgCl_2$	1.5	1.5 mM
Taq DNA polymerase (5U/μl)	0.25	.5 units
Primer A (10 μm)	2.5	0.5 μM
Template DNA	2.5	15-20ng
Distilled sterile water	37.25	—

quantities in the PCR reaction mixture compared to the template DNA, chances of formation of primer-template complexes at the primer annealing sites are more upon lowering the temperature. Thus annealing of primers on the DNA strands allows the third step of PCR to begin. In this step, annealed primers start polymerization/synthesis of new strands with the aid of DNA polymerase enzyme and associated salts in PCR buffer. Hence, exponential amplification of target sequences is obtained after repeating these three- step cycles 25-40 repeated times. PCR amplification of specific DNA fragments is influenced by the temperature profile of PCR machines, primer annealing temperature, amount of template DNA and quality of other associated chemicals. The non - specific hybridization of primers can be avoided by keeping high annealing temperature. Though specific, semispecific and arbitrary primers have been used for the PCR, in DNA fingerprinting, use of short oligonucleotide arbitrary primers is more common which are available from different companies on custom synthesis basis.

Table 2. Standard temperature profiles used for PCR

PCR Cycle 1:	Denaturation of genomic DNA 94	°C for 5 min.
	Primer annealing	40-60 °C for 1 min
	Amplification	72 °C for 1 min.
Forty PCR cycles:		95 °C for 1 min.
		40-60 °C for 1 min.
		72 °C for 2 min.
Last cycle:		72 °C for 5 min.

Once PCR reactions are completed, the PCR products are resolved by agarose gel electrophoresis and staining with ethidium bromide.

Advantages

This technique allows the specific and sensitive detection and amplification of DNA fragments from the nucleic acids obtained from a variety of materials like blood stains and animal/plant

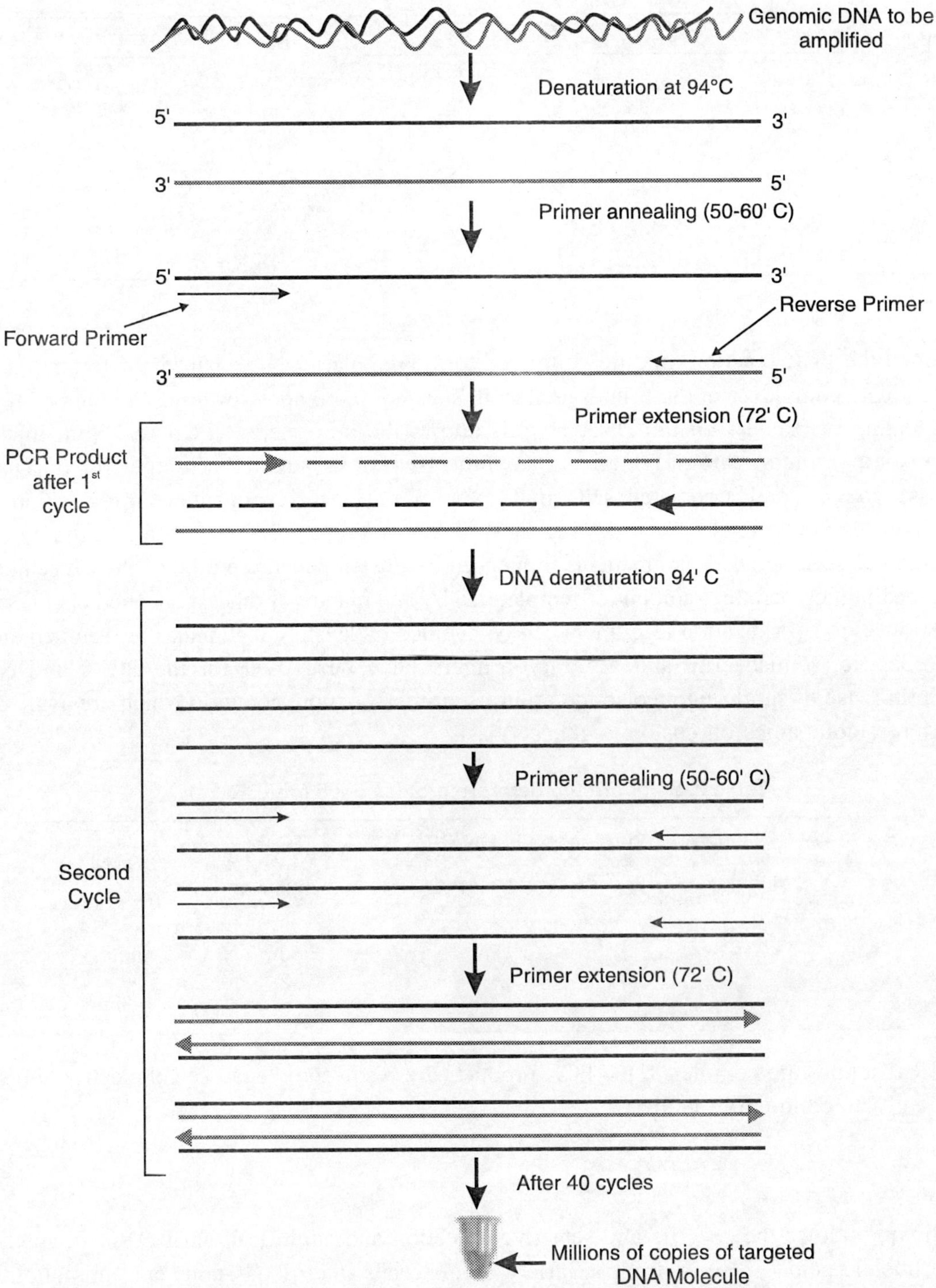

Fig. 1. Different steps of polymerase chain reaction.

fossil DNA. More importantly, a very minute quantity of DNA is sufficient to start the PCR reaction.

PCR Applications

The PCR techniques have unlimited applications in biotechnology. Some of them are listed below:

- DNA fingerprinting of different organisms.
- Development of molecular maps of different species.
- Molecular tagging and mapping of importance genes.
- Gene pyramiding by using marker assisted selection.
- PCR cloning of important genes and promoters.
- Automated cyclic sequencing of DNA.
- Identification of inserted genes in the target organisms.
- Expression analysis of genes using quantitative PCR.

DESIGNING PCR PRIMERS

What is primer?

Primer is a short stretch of oligonucleotide which prime or initiate the synthesis of new DNA strand in typical PCR reaction. The primers are now essential components of any molecular biology experiment whether it is related to amplification of a gene, fingerprinting genomes, sequencing a specific genome fragments and in gene expression studies.

Criteria used for primer design

Though, computationally it is very simple to design a PCR primer, it has to follow specific criteria. The general criteria for the design of primers are:

i) The minimum length of the primer should be of 15 nucleotides or and optimum 20-22 nt for more specific amplification.

ii) The GC content should vary between 40-60%.

iii) The Tm (melting temperature) should be between 50-60° C.

iv) Tm for both forward and reverse primer should be as close as possible.

v) The predicted PCR product should by atleast between 100-150bp.

There are some thumb rules which one should remember while designing primers for any amplification. The longer the primer, having more GC contents and high Annealing temperature, it will avoid non-specific amplifications from the target DNA. The annealing temperature of a primer should be less than 5°C from its Tm. Besides that, self annealing and formation of primer dimer should be avoided. The primer binding sites should be unique so that a specific fragment is amplified. Therefore, after designing primers, one should perform BLAST search against the genome databases of the organism to find out its uniqueness.

EXERCISE

Use of primer design software

Many primer design software are available on world wide web. Therefore one has to be very careful while choosing specific primer design software. Best way would be to look at the literature and find out the maximally used software tool for designing primers in molecular biology experiments. It will give an idea that the software is really good and primer designed by this are successfully validated experimentally. In most of the molecular biology reports Primer3 software freely available in the public domain on http://frodo.wi.mit.edu/ has been extensively used. Since it could be accessible from any desktop connected to the internet which has been demonstrated in this chapter. Primer 3 software picks up primers based on the criteria mentioned earlier like, primer length, Tm, GC contents, avoid primer dimer and PCR products along with some miscellaneous criteria. In most of the cases, default setting works fine. However, user can select the criteria from the drop down boxes and pick primers according to the requirement of an experiment.

Example: Design the primer pair for the following DNA sequence using Primer3 software.

Step 1. Select DNA sequence for primer design. An example of the sequence is given in Fig. S1.

Step 2. Open PRIMER3 home page. (http://frodo.wi.mit.edu/) (Fig. S2).

Step 3. Paste the sequence in software window (Fig. S3). Alternatively the sequence file can be browsed from its name or from any location from the desktop. Select different primer design parameters. Generally, we use the following parameters for primer designing.

a. GC content = 40-60

b. Primer length = 20-22

c. Product size = 150-200

d. 3′ end = G

```
ccaagcttccaagtcaatgaggcaaataaatgtgtatactatcactatggtaggggtgaggg
agttattctgtgcttgtatgttgatgacatactaatatttgagacaaactaccgaggtgactaaa
ctgacggttaagatcattctgtctcaacagctctctcgactatgaacgaagcttgggagctacg
ctgactgcttactacctaacacccttacggaaacacatagagcgtgaccacactgggacttaa
aactctatgccaagcatcccactctactggtgaacgaacgatcttgactcgcctcacggccat
accccttgacaggccaacgccatttccctccccaccataatccttcaagtaaattcccctaacgc
cttcaggctgccctaccgcccaacagaaaccccaacccaagacacttcggcctaacgcgaaa
ctccacacactcggacgactaattcccccccaacacgccaattcaagaaaccactaaactgg
tcatgagctggcctgacaccgctcgggacgcggtccgtcaccacccacctaggcgcttcgga
agcgtagaagccccactaacccccactggtgacgctccacacacacatacgcggctaaacca
ccacgtcgaaatggtggaatacgcggacaacatcccgctgtcaacttatgactcctgtcgcat
gacccagtccaggaagcattaaacc
```

Fig. S1. Target nucleotide sequence to be used for primer design.

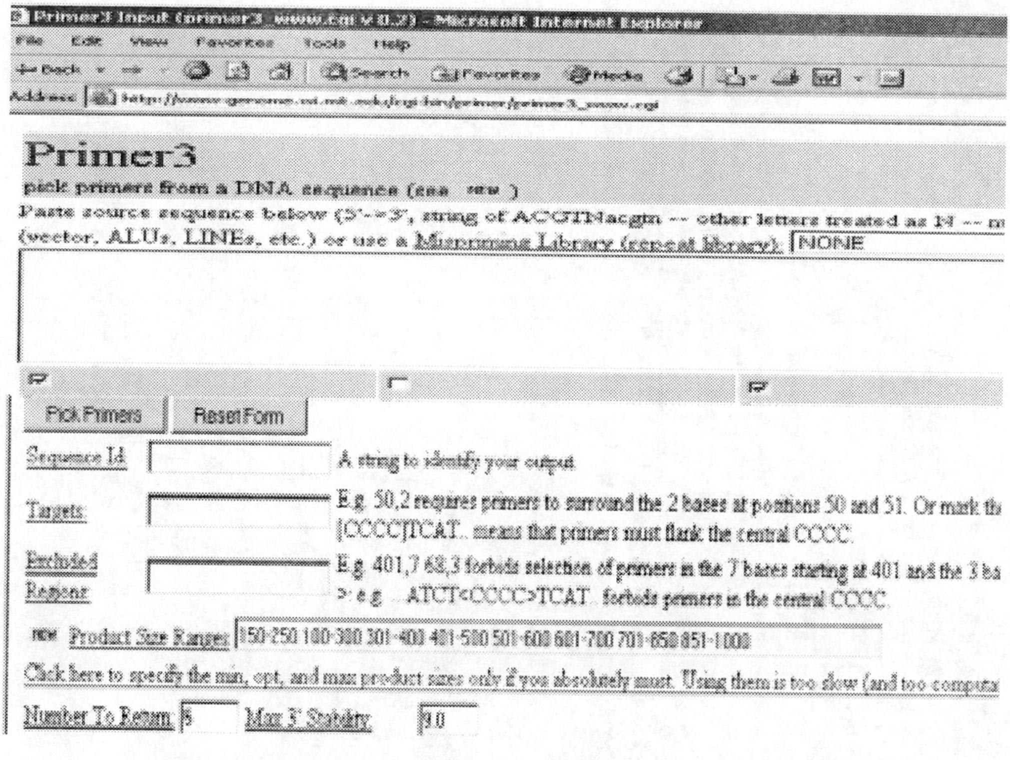

Fig. S2. Main window of primer3 software.

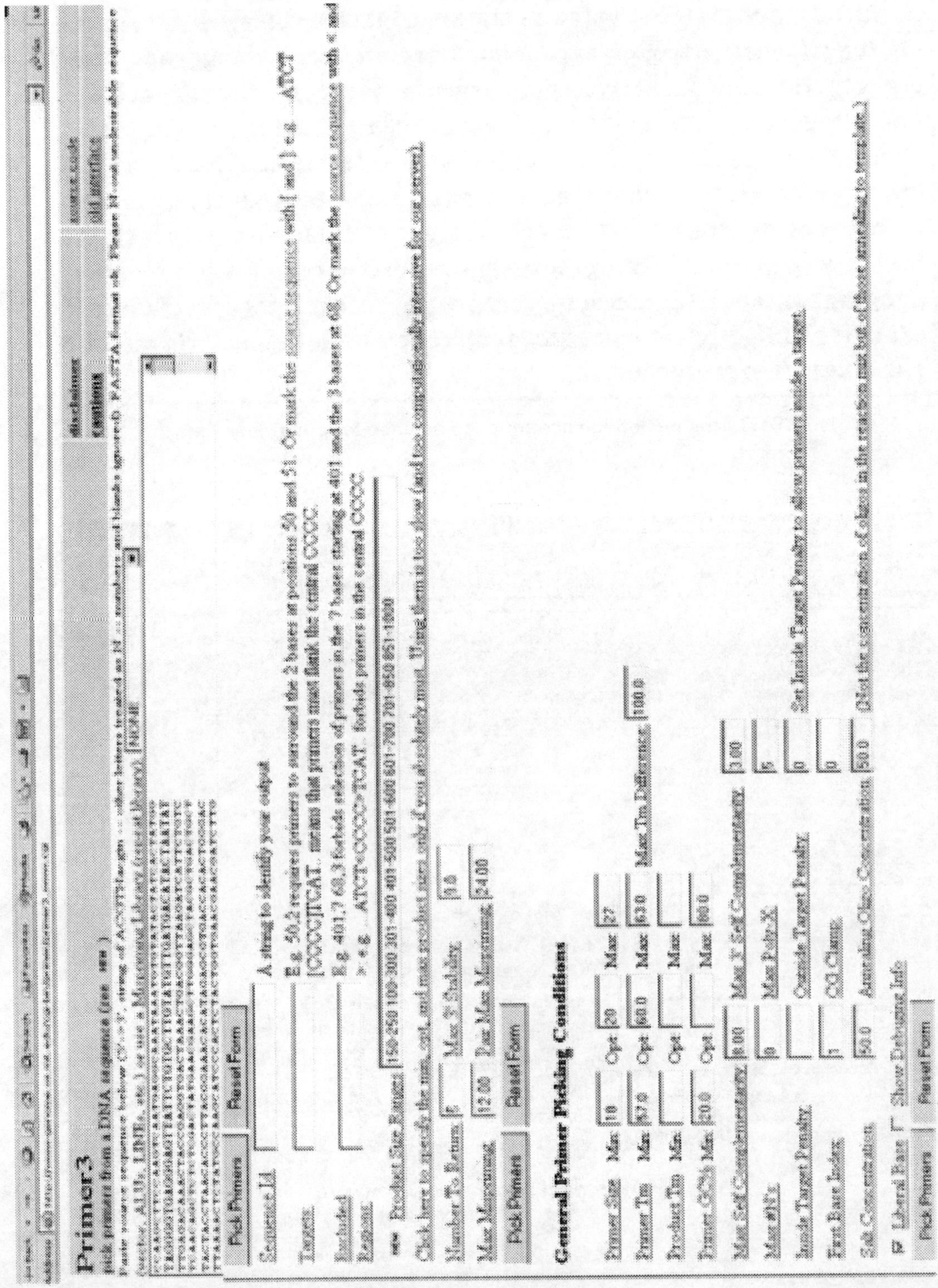

Fig. S3. The targets sequence has been pasted in the window.

Step 4. After setting the parameters click on **Pick Primers**. After some time, result page will be opened having information about forward and reverse primers, their length, melting temperature, CG content and their position in the sequence (Fig. S4).

Primer3 Output

```
No mispriming library specified
Using 1-based sequence positions
OLIGO            start  len     tm     gc%    any    3'  seq
LEFT PRIMER        219   20  59.86   50.00   3.00  1.00 CGGAAACACATAGAGCGTGA
RIGHT PRIMER       590   20  59.83   60.00   3.00  0.00 GGGGTTAGTGGGGCTTCTAC
SEQUENCE SIZE: 716
INCLUDED REGION SIZE: 716

PRODUCT SIZE: 372, PAIR ANY COMPL: 4.00, PAIR 3' COMPL: 2.00

    1 CCAAGCTTCCAAGTCAATGAGGCAAATAAATGTGTATACTATCACTATGGTAGGGGTGAG

   61 GGAGTTATTCTGTGCTTGTATGTTGATGACATACTAATATTTGAGACAAACTACCGAGGT

  121 GACTAAACTGACGGTTAAGATCATTCTGTCTCAACAGCTCTCTCGACTATGAACGAAGCT

  181 TGGGAGCTACGCTGACTGCTTACTACCTAACACCCTTACGGAAACACATAGAGCGTGACC
                                          >>>>>>>>>>>>>>>>>>>>
  241 ACACTGGGACTTAAAACTCTATGCCAAGCATCCCACTCTACTGGTGAACGAACGATCTTG

  301 ACTCGCCTCACGGCCATACCCTTGACAGGCCAACGCCATTTCCCTCCCCACCATAATCCT

  361 TCAAGTAAATTCCCCTAACGCCTTCAGGCTGCCCTACCGCCCAACAGAAACCCCAACCCA

  421 AGACACTTCGGCCTAACGCGAAACTCCACACACTCGGACGACTAATTCCCCCCCAACACG

  481 CCAATTCAAGAAACCACTAAACTGGTCATGAGCTGGCCTGACACCGCTCGGGACGCGGTC

  541 CGTCACCACCCACCTAGGGGCTTCGGAAGCGTAGAAGCCCCACTAACCCCACTGGTGACG
                                  <<<<<<<<<<<<<<<<<<<<
  601 CTCCACACACACATACGCGGCTAAACCACCACGTCGAAATGGTGGAATACGCGGACAACA

  661 TCCCGCTGTCAACTTATGACTCCTGTCGCATGACCCAGTCCAGGAAGCATTAAACC

KEYS (in order of precedence):
>>>>>> left primer
<<<<<< right primer

ADDITIONAL OLIGOS
                   start  len     tm     gc%    any    3'  seq
 1 LEFT PRIMER       219   20  59.86   50.00   3.00  1.00 CGGAAACACATAGAGCGTGA
   RIGHT PRIMER      616   20  59.73   60.00   3.00  2.00 TATGTGTGTGTGGAGCGTCA
   PRODUCT SIZE: 397, PAIR ANY COMPL: 7.00, PAIR 3' COMPL: 3.00

 2 LEFT PRIMER       219   20  59.86   50.00   3.00  1.00 CGGAAACACATAGAGCGTGA
   RIGHT PRIMER      599   20  60.28   60.00   5.00  0.00 GTCACCAGTGGGGTTAGTGG
```

Fig. S4. The output window of primer3 software showing details of the primers.

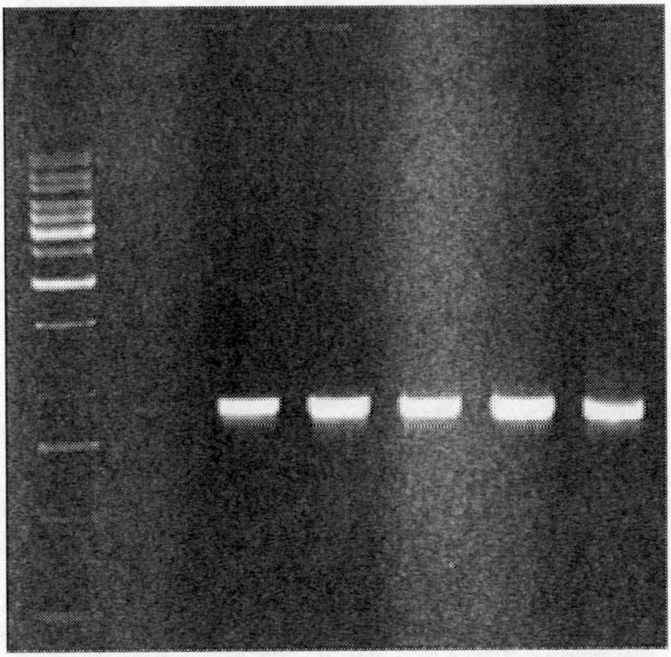

Fig. S5. Allele specific PCR products amplified from rice plants (Lanes 2-7) using primer combinations designed with Primer 3 software. DNA size marker is given in lane 1.

Step 5. Use primer in a typical PCR reaction (Table 1) for the amplification of specific alleles. The PCR products are run on an agarose gel, stained with ethidium bromide and photographed on a gel documentation system as shown in the photographs (Fig. S5).

Primer 3 software gives a best pair combination for the given sequence along with GC contents, TM, length and product size at the top. Beside this it also gives four more options to select if first option does not work in the experiment.

Suggested Readings

Saiki R. K., D. H. Gelfand, S. Stiffel, R. H. Higuchi, G. T. Horn, K. B. Mullis, and H. A. Erlich (1985), Primer directed enzymatic amplification of DNA with a thermostable DNA polymerase, *Science*, **239**: 487-492.

Lincoln, S., M Daly and M. E.S. Lander (1991), PRIMER: a computer program for automatically selecting PCR primer. The Whitehead Institute. Freely available from the authors (http://www-genome.wi.mit.edu/genome_software/other/primer3.html)

Rozen, S. and H.J. Skaletsky (2000), Primer3 on www for general users and for biologist programmers. *Methods Mol. Biolo.* **132:** 365-386.

1 Appendix

Introduction to Basic Softwares used in Bioinformatics

Biological data are being generated from different species world over. One of the most challenging tasks is to randle store and analyse this huge amount of data. Therefore, for computational analysis of the data some of the basic software are essentially required. Introduction to these software, particularly to the biologists would be of great help in understanding the sequence data management and analysis.

SOFTWARE USED FOR DATABASE DEVELOPMENT

Storage, management and sharing of genome sequence data among the end users over world wide web are essential for which there is a need to develop biological databases. For the development of intelligent biological databases, a Database Management System (DBMS) is required. The most commonly used DBMS have been explained below:

Oracle

The Oracle Database is a **R**elational **D**atabase **M**anagement **S**ystem (RDBMS) produced and marketed by Oracle Corporation (www.oracle.com/technology/database/index.html). The Oracle is a DBMS which is based on the relational model in which data is stored in the form of tables, fields and records. The latest version of Oracle database is 11g (g-grid). It is very fast in data handling and has unique ability to deliver grid computing. It is a paid software. The Oracle Corporation offers term licensing for all Oracle products.

Uses

- Oracle allows to store, and access vast volumes of data with greater speed of the servers. It is being used in the management of several important biological databases throughout the world.
- Oracle is much versatile than other RDBMS. Oracle can run and handle more transactions compared to other RDBMS making data access and application software with faster speed.
- Oracle is very useful for creating large databases and also access of databases by multiple users can be obtained with greater speed.
- It has sophisticated features, highly flexible and can run on multiplatform.

MySQL

MySQL is one of the most popular Open Source SQL database management systems. Any one can download, modify and use it freely. It can be downloaded from www.mysql.com. The SQL stands for "Structured Query Language" which is the most common language used to access data from the databases and qualities to the ANSI/ISO SQL standards. The MySQL is developed distributed, and supported by MySQL AB (www.mysql.com) and written in C and C++. It is also a relational DBMS that stores data in separate tables and files. This software offers a rich and useful source of functions which is highly suited for the databases accessed on the world wide web. Currently MySQL5.0 version is available for free downloads.

Uses

- One of the most widely used freely available RDBMS which is popular, flexible and robust.
- Being free ware, it is being used for the development of most of the public databases.
- No recurring cost is involved with this software since the new versions can be freely downloaded for up gradations.
- MySQL DBMS is available along with the source codes, which can be changed and modified as per the requirement of the users.

COMPUTER PROGRAMMING LANGUAGES

C Language

An important computer programming language "C" was developed by Dennis Ritchie (1972) at the Bell Telephone Laboratories. This general purpose, procedural, cross-platform and block structured language is used with the UNIX operating system, different software platforms and computer architectures. It was named as "C" because its many features have been derived from the earlier version "B". "C" is generally called "Middle Level" programming language and is widely used for developing application software. Developing a software programme in 'C' often requires four steps like editing, compiling, linking and executing. Other higher-level languages also use the compilers, libraries and interpreters of 'C' for their implementation. Specific features of "C" includes, use of pointers for memory, array, structures and functions. Besides, it can produce efficient programmes compiled on a variety of computers. The 'C' language can be used in operating system, language compiler, text editors, assemblers, point spoolers, Network Drivers databases and language interpreters.

Uses

- It can work on multiplatform computers.
- Widely used for developing application software.
- It is widely used to implement end-user applications.
- Due to code portability and efficiency, it was used in operating and embedded system applications.
- In bioinformatics, it can be used in algorithms for parsing BLAST outputs files etc.

C++ LANGUAGE

The extended form of the 'C' language named as C++ was developed by Bjarne Stroustrup with many attractive features. These features are efficiency, closeness to the machine, and a variety of built-in types. A number of new features added to C++ have to make it more robust programming language. The OOP (Object Oriented Programming) features have been added to 'C' language so that individual programmer can enjoy writing good programs in easier way. A language is called OOP, if it has features like abstraction, encapsulation, inheritance, and polymorphism. The hiding of information is called encapsulation. Encapsulation is implemented in C++ by allowing all members of a class to be declared as either public, private, or protected. As per the object oriented principle, the type definition should encapsulate all of the functions that access the internal representation of a type. The inheritance means acquiring the properties of one data type by other data types. If an operator or function uses its ability in different ways, then it is called polymorphism. Polymorphism is a powerful feature of the object oriented programming language C++. Other important features of C++ are constructors and destructors. The purpose of a constructor is to initialize an object whereas the destructors are used to cleanup and deallocate memory for a class object and its class members when the object is destroyed.

Uses

- Very robust programming language.
- Very useful language for writing bioinformtaics software due to a 'C' compiler which helps it running on multicore environment.
- A very fast computational language.
- Small programme written in C++ can handle large amount of the data.

PERL (Practice Extraction and Reporting Language)

With the increasing importance of bioinformatics, the importance of computer programming in biology has also increased tremendously. Many programming languages such as C++, JAVA, Python and Perl etc. are being used widely in Bioinformatics. Perl is a popular programming language that is extensively used in Bioinformatics and web programming. Being a scripting language, it is very helpful in solving simple biological problems like finding reverse complements of DNA, DNA to protein conversion, motif finding, gene finding, sequence assembly and format conversions. It has the ability to handle long strings of DNA or RNA and perform analysis on them. The current version of the Perl is Perl v-5.10.0 which is freely available on the website www.perl.com and runs on all operating systems including Microsoft Windows, UNIX, Linux, Macintosh etc. Perl application must be installed on the computer to run the Perl program. The computer language should have a translator application (compiler) that can turn programs into the instructions. The instructions are then given to the computer which can execute and give the output. Perl application is often referred as the Perl interpreter which also includes a compiler. Different styles of writing scripts such as Imperative programming, Object Oriented Programming, Functional programming and Logical programming etc are provided within the Perl.

BioPerl

Bioperl is the bundle of more than 500 different modules of Perl scripting language which are extensively used in bioinformatics research. An international group of volunteers is writing and maintaining these modules. Therefore, BioPerl is an open source, object oriented Perl modules or a bioinformatics Toolkit used for format conversion, report processing, data manipulation, sequence analysis and batch processing etc. The latest version of BioPerl is 1.4 which can be freely downloaded from http://bioperl.org.

Uses of Per and BioPerl

- The programs can be written very easily in Perl since it is a scripting language. It simplifies several common bioinformatics tasks. The long DNA and amino acid sequences can be processed and manipulated very easily with Perl. The ASCII text files or Flat files which are used in GenBank, PDB and other biological databases can be easily read by Perl.

- Several bioinformatics applications can be automated by using Perl.
- It is a very fast language hence provides speed to the biological research. A small script written in Perl can be used to solve complex biological problems.
- Portability is another very useful aspect of Perl. The programs written in Perl on any of the operating system can be run on different operating systems.
- It can easily and efficiently handle huge amount of biological data.
- It doesn't require any Internet access.
- Bioperl modules are readymade modules which can be used in a very simplified manner to get desired output.

JAVA

The Programming language Java has been developed as a core component of Sun Microsystems in 1995. Though many of its syntax derived from C and C++, it has fewer low level facilities and a simple object model. It also combines the syntax for generic, structured and object oriented programming. The byte code is used to java applications which can run on any Java virtual machine (JVM) on multiple computer architecture. Java is basically a combination of three important things.

- A high-level, object-oriented Java programming language.
- A high-performance Java Virtual Machine (JVM) in which bytecodes are executed on a specific computing platform.
- A Java platform, where Java bytecodes are compiled by running JVM on a set of standard library e.g. provided in the Java Standard Edition (SE) or Enterprise Edition (EE).

The most important feature of Java is that it is platform independent and can run on Microsoft Windows, Linux, and Unix platforms. The current release of java is Java 6 version1.6x which can be freely downloaded from www.sun.com.

Uses

- It is an open source language.
- Computational platform independent.
- Programs are easy to write and debug in java.
- Use object oriented language with several characteristic features.

- Code is compiled to bytecodes that are easily interpreted by Java virtual machines.
- For compilation and interpretation, two steps are used which improves security by extensive code checking.
- It provides extensive front-end to softwares and public databases in the form of Java Server Pages.
- Java is popular for its security and network awareness.
- Java-based data retrieval methods for accessing genomic sequence and annotation are very useful.
- The transfer of data between different data storage standards can be facilitated by BioJava
- Java tools can be effectively used for processing biological data, manipulating sequences, dynamic programming, file parsers and simple statistical analyses.

2 Appendix

Genome size of important organisms which have been sequenced

Organism	Scientific name	Importance	Genome Size (kb)	Database
Archaea	*Ignicoccus hospitalis*	Biotechnological	1434	Joint Genome Institute
Archaea	*Methanocorpusculum labreanum*	Biotechnological, Energy Production	1739	Joint Genome Institute
Archaea	*Aeropyrum pernix*	Biotechnological	1700	NITE
Archaea	*Pyrococcus horikoshii (shinkaj)*	Biotechnological	1955	Univ of Tokyo, NITE
Archaea	*Archaeoglobus fulgidus*	Biotechnological	2420	J. Craig Venter Institute, Univ of Illinois at Urbana-Champaign
Archaea	*Methanothermobacter thermoautotrophicus*	Biotechnological, Energy production	1873	Genome Therapeutics, Ohio State Univ
Archaea	*Methanocaldococcus*	Biotechnological, Energy production	1729	J. Craig Venter Institute, Univ of illinois at Urbana-Champaign Center of Bioengineering
Bacteria	*Yersinia*	Human Pathogen, Medical	4192	Los Alamos National Labora tory, Joint Genome Institute
Bacteria	*Acinetobacter*	Human Pathogen, Medical	3607	Genoscope
Bacteria	*Xanthomonas oryzae pv. oryzae*	Agricultural, Plant Pathogen	5083	J. Craig Venter Institute, Univ of Maryland-CDCB
Bacteria	*Clostridium botulinum*	Human Pathogen, Medical	3256	Los Alamos National Labora tory, Joint Genome Institute
Bacteria	*Clostridium botulinum B*	Medical	3473	Los Alamos National Labora tory, Joint Genome Institute
Bacteria	*Salinispora arenicola*	Biotechnological, Cancer treatement, Medical	4917	Joint Genome Institute
Bacteria	*Azorhizobium caulinodans*	Agricultural	4717	Univ of Tokyo

Organism	Scientific name	Importance	Genome Size (kb)	Database
Bacteria	*Arcobacter but zleri*	Human Pathogens, Medical	2259	ARS-USDA, Agencourt Bioscience
Bacteria	*Bacillus pumilus*	Bioenergy, Biotechnological, Medical	3681	BCM-HGSC
Bacteria	*Serratia Proteamaculans*	Medical, Human Pathogen, Environmental, Bioremediation	4891	Joint Genome Institute
Bacteria	*Shewanella sediminis*	Biotechnological, Degrades RDX	4497	Joint Genome Institute
Bacteria	*Campylobacter concisus*	Medical, Human Pathogen	1929	J. Craig Venter Institute
Bacteria	*Streptococcus gordonii*	Medical, Human Pathogen	2051	J. Craig Venter Institute
Bacteria	*Citrobacter koseri*	Medical, Animal Pathogen, Human Pathogen	5003	Washington Univ
Bacteria	*Escherichid coli O139:H28*	Medical, Human Pathogen	4755	J. Craig Venter Institute
Bacteria	*Escherichia coli O9*	Medical, Human Pathogen	4384	J. Craig Venter Institute
Bacteria	*Staphylococcus aureus*	Animal Pathogen, Cattle Pathogen, Human Pathogen, Medical, Poultry Pathogen	2698	Juntendo Univ
Bacteria	*Enterobacter sakazakii*	Medical, Human Pathogen	4277	Washington Univ
Bacteria	*Vibro harveyi*	Environmental	5944	Princeton Univ, Washington Univ
Bacteria	*Francisella tularensis holarctica*	Medical, Human Pathogen, Biothreat	2079	Los Alamos National Laboratory, Joint Genome Institute, Biohealthbase
Bacteria	*Bacillus amyloliquefaciens*	Biotechnological, Antibiotic production, Suppresses Plant Pathogens, Agricultural	3693	Competence Network Goettingen Genomics Laboratory
Bacteria	*Pseudomonas aeruginosa*	Medical, Human Pathogen, Animal Pathogen, Plant Pathogen, Agricultural	6286	J. Craig Venter Institute
Bacteria	*Actinobacillus succinogenes*	Biotechnological, Succinic-acid production	2079	Joint Genome Institute
Bacteria	*Klebsiella pneunomiae*	Animal Pathogen, Cattle Pathogen, Human Pathogen, Medical	4776	Washington Univ
Bacteria	*Staphylococcus aureus*	Animal Pathogen, Cattle Pathogen, Human Pathogen, Medical, Poultry Pathogen	2747	Joint Genome Institute, Rockerfeller Univ
Bacteria	*Clostridium beijerinckii*	Energy production, Solvent production, Biotechnological	5020	Joint Genome Institute, Univ of Illinois
Bacteria	*Bacteroides vulgatus*	Medical, Human Pathogen, Animal Pathogen, Human Pathogen, medical	4065	Washington Univ

Organism	Scientific name	Importance	Genome Size (kb)	Database
Bacteria	*Lactobacillus reuteri*	Food industry, Biotechnological	1900	Joint Genome Institute, Univ of Otago
Bacteria	*Mycoplasma agalactiae*	Medical, Animal Pathogen	742	Genoscope
Bacteria	*Clostridium botulinum*	Medical, Human Pathogen, Biothreat	3574 Reading	Sanger Institute, Institute of Food Research. Univ of
Bacteria	*Staphylococcus aureus*	Medical, Human Pathogen, Animal Pathogen, Cattle Pathogen, Poultry Pathogen,	2697	Joint Genome Institute Rockefeller Univ
Bacteria	*Clavibacter michiganensis*	Agricultural, Plant Pathogen	2984	Bielefeld Univ
Bacteria	*Bradyrhizobium sp*	Agricultural	7394	Joint Genome Institute
Bacteria	*Vibrio cholerae*	Medical, Human Pathogen	3875	J. Craig Venter Institute
Bacteria	*Bradyhizobium sp*	Agricultural	6717	Genoscope
Bacteria	*Rhodobacter*	Bioremediation, Energy production, Biotechnological, Environmental	3111	Joint Genome Institute
Bacteria	*Streptococcus suis*	Medical, Human Pathogen, Swine Pathogen, Animal Pathogen	2189	Beijing Institute of Genomics
Bacteria	*Streptococcus suis*	Medical, Human Pathogen, Swine Pathogen, Animal Pathogen	2186	Beining Institute of Genomics
Bacteria	*Enterobacter sp*	Agricultural	4115	Joint Genome Institute
Bacteria	*Pseudomonas stutzeri*	Environmental, Agricultural, Bioremediation	4128	Chinese Academy of Agricultural sciences
Bacteria	*Yersinia Pestis*	Medical, Human Pathogen, Animal Pathogen, Biothreat	3850	Joint Genome Institute
Bacteria	*Mycobacterium flavenscens (gilvum)*	Biotechnological	5241	Joint Genome Institute
Bacteria	*Burkholderia vietnamiensis*	Medical, Animal Pathogen, Plant Pathogen, Human Pathogen, Agricultural	7617	Joint Genome Institute
Bacteria	*Clostridium thermocellum*	Biotechnological, Energy production	3191	Joint Genome Institute, Univ of Rochester
Bacteria	*Streptococcus sanguinis*	Medical, Human Pathogen	2270	Commonwealth Biotechnologies, Inc, Virginia Commonwealth Univ
Bacteria *cremoris*	*Lactococcus lactis* Biotechnological	Food Industry,	2434	National Univ of Ireland, Alimentary Pharmabiotic Centre, Institute of Food Research, Bielefeld Univ, Univ of Groningen

Organism	Scientific name	Importance	Genome Size (kb)	Database
Bacteria	*Burkholderia mallei*	Medical, Biothreat, Human Pathogen, Animal Pathogen	5510	J. Craig Venter Institute
Bacteria	*Bartonella bacilliformis*	Medical, Human Pathogen	1283	J. Craig Venter Institute
Bacteria	*Escherichia coli O1: K1:H7*	Avian Pathogen, Medical	4467	Iowa State Univ
Bacteria	*Mycobacterium ulcerans*	Medical, Human Pathogen	4160	Institute Pasteur
Bacteria	*Bacillus thuringiensis*	Agricultural, Biotechnological, Insect Pathogen	4736	Los Alamos National Laboratory, Joint Genome Institute
Bacteria	*Mycobaterium avium*	Medical, Animal Pathogen, Human Pathogen	5120	J. Craig Venter Institute
Bacteria	*Streptococcus pyogenes*	Medical, Human Pathogen	1745	Sanger Institute, Univ of Newcaste
Bacteria	*Shigella flexneri 5b*	Medical, Biothreat, Human Pathogen	4116	Microbial Genome Center, Beijing
Bacteria	*Rhizobium leguminosarum*	Biotechnological, Agricultural	4700	Sanger Institute, Univ of East Anglia. Univ of York
Bacteria	*Burkholderia xenovorans (fungorum)*	Bioremediation, Environmental, Medical, Human Pathogen, Plant Pathogen, Agricultural	8702	Joint Genome Institute, Michigan State Univ
Bacteria	*Escherichia coli*	Human Pathogen, Human Gut Microbiome Initiative (HGMI), Medical	5044	Washington Univ
Bacteria	*Rickettsia bellii*	Agricultural	1469	CNRS
Bacteria	*Ehrilichia chaffeensis*	Medical, Human Pathogen	1105	Ohio State Univ, J. Craig Venter Institute
Bacteria	*Staphylococcus*	Animal Pathogen, Cattle Pathogen, Human Pathogen, Medical, Poultry Pathogen	2892	Univ of Oklahoma
Bacteria	*Frankia sp*	Agricultural, Biotechnological	4499	LBNL, Joint Genome Institute, Univ of Connecticut, Univ of New Hampshire
Bacteria	*Xanthomonas*	Agricultural, Plant Pathogen	4372	NIAS, Japan
Bacteria	*Burkholderia thailandensis*	Agricultural	5645	J. Craig Venter Institute
Bacteria	*Pseudomonas syrinagae phaseolicola*	Agricultural, Plant Pathogen	4984	Cornell Univ, J. Craig Venter Institute
Bacteria	*Xanthomonas campestris vesicatoria*	Agricultural, Plant Pathogen	4487	Biefield Univ
Bacteria	*Xanthomonas campestris*	Agricultural, Plant Pathogen	4273	The Institute of microbiology, China, Guangxi Univ
Bacteria	*Pseudomonas syringae*	Agricultural, Plant Pathogen	5090	Joint Genome Institute, Univ of California, Berkeley

Organism	Scientific name	Importance	Genome Size (kb)	Database
Bacteria	*Brucella abortus*	Medical, Human Pathogen, Animal Pathogen, Cattle Pathogen, Biothreat	3085	Univ of minnesota, Univ of Uppsala, IIB-UNSAM
Bacteria	*Xanthomonas oryzae*	Agricultural, Plant Pathogen *pv oryzae*	4637	MACROGEN, NIAB
Bacteria	*Leifsonia xyli*	Plant Pathogen, Agricultural	2030	Univ of Compinas
Bacteria	*Bacillus thuringiensis konkukian*	Biotechnological, Agricultural, Insect Pathogen	5117	Los Alamos National Laboratory Joint Genome Institute
Bacteria	*Themus thermophilus*	Biotechnological	1982	Goettingen Genomics Laboratory
Bacteria	*Mycoplasma mycoides SC*	Cattle Pathogen, Animal Pathogen, Medical	1016	Biefield Univ, National Verterinary Inst, Uppsala, Royal Institute of Technology, Stockholm
Bacteria	*Candidatus Pytoplasma onion yellows*	Plant Pathogen, Agricultural	754	Univ of Tokyo
Bacteria	*Pseudomonas syringae tomato*	Agricultural, Plant Pathogen	5470	Univ of Nebraska, Kansas State Univ, Cornell Univ, J. Craig Venter Institute, Univ of Missouri
Bacteria	*Mycoplasma gellisepticum*	Medical, Food industry, Poultry Pathogen, Animal Pathogen	726	Unive of Connecticut
Bacteria	*Helicobacter hepaticus*	Medical, Mice Pathogen, Animal Pathogen	1875	MWG-Biotech, MIT, Univ of Wuerzburg, GeneData
Bacteria	*Rickettsia sibirica*	Medical, Human Pathogen	1234	Univ of Maryland School of Medicine, CDC, Agencourt Bioscience
Bacteria	*Clostridium tetani*	Medical, Human Pathogen, Animal Pathogen	2373	Goettingen Genomics Laboratory
Bacteria	*Xylella Fastidiosa-grape*	Agricultural, Plant Pathogen	2034	AEG Branilian Consotium
Bacteria	*Bradyrhizobium japanicum*	Agricultural	8317	Kazusa DNA Research Institute
Bacteria	*Escherichia coli O6:K2:H1*	Medical, Human Pathogen	5379	Univ of Wisconsin
Bacteria	*Xanthomonas axonopodis pv. cotri*	Agricultural, Plant Pathogen	4312	FAPESP, Univ of Compinas, Univ of Sao Paulo
Bacteria	*Xanthomonas campestris*	Agricultural, Plant Pathogen	4181	FAPESP, Univ of Sao Paulo
Bacteria	*Ralstonia solanacearum*	Agricultural, Plant Pathogen	3440	CNRS, INRA, Genoscope
Bacteria	*Agrobacterium tumefaciens*	Agricultural, Plant Pathogen	5402	DuPont, Univ of Washington, Univ of Campinas
Bacteria	*Agrobacterium tumefaciencs*	Agricultural, Plant Pathogen	4548	Univ of Richmond, Monsanto, Cereon Genomics

Organism	Scientific name	Importance	Genome Size (kb)	Database
Bacteria	*Mesorhizobium loti*	Environmental, Bioremediation	6743	Kazusa DNA Research 0Institute
Bacteria	*Xylella fastidiosa CVC*	Agricultural, Plant Pathogen	2766	ONSA
Bacteria	*Chlamydophila pneumoniae*	Medical, Human Pathogen	1069	Yamaguchi Univ, Kyushu Univ, RIKEN
Bacteria	*Bacillus holodurans*	Biotechnological, Alkaliphilic enzyme production, Loundry detergents	4066	JAMSTEC
Bacteria	*Neisseria meningitidis*	Medical, Human Pathogen	2065	Sanger Institute, Max Planck Institute, Univ of Oxford
Bacteria	*Thermotoga maritima*	Biotechnological, Energy production, Evolutionary	1858	J. Craig Venter Institute
Bacteria	*Chlamydophila*	Medical, Human Pathogen *pneumoniae*	1052	Stanford Univ, Univ of California, Berkeley
Bacteria	*Helicobacter pylori*	Medical Human Pathogen	1491	Astra, Genome Therapeutics
Bacteria	*Ricksttsia prowazekii*	Medical, Biothreat, Human Pathogen	835	Univ of Uppsala
Bacteria	*Chlamydia trachomatis*	Animal Pathogen, Human Pathogen, Medical	895	Stanfor Univ, Univ of California, Berkeley
Bacteria	*Treponema pallidum*	Medical, Human Pathogen	1036	BCM-HGSC, Univ of Texas, J. Craig Venter Institute
Bacteria	*Mycobacterium tuberculosis*	Medical, Human Pathogen, Animal Pathogen	4402	Sanger Institute
Bacteria	*Aquifex aeolicus*	Biotechnological	1529	Univ of Illoinois at Urbana-Champaign, Diversa
Bacteria	*Borrelia burgdorferi*	Medical, Human Pathogen	851	Brookhaven Natl Lab, J. Craig Venter Institute
Bacteria	*Bacillus subtilis*	Biotechnological	4105	Jananese Consortium, European Consortium
Bacteria	*Helicobacter pylori*	Medical, Human Pathogen	1576	J. Craig Venter Institute
Bacteria	*Mycoplasma pneumoniae*	Medical, Human Pathogen	689	Univ of Heidelberg
Bacteria	*Candidatus Phytoplasma australiense*	Agricultural	840	Max Planck Institute, Charles Darwin Univ
Bacteria	*Borrelia hermsii*	Human Pathogen, Medical	819	RML-NIAID
Bacteria	*Lactobacillus reuteri*	Food industry, Biotechnological	1820	Univ of Tokyo, Azabu Univ
Bacteria	*Lactobacillus fermentum*	Biotechnological, Food industry	1843	Univ of Tokyo, Azabu Univ
Bacteria	*Vibrio fischeri*	Environmental, Marine Microbial Initative (MMI)	3844	Univ of Georgia, J. Craig Venter Institute
Bacteria	*Chlamydia trachomatis*	Animal Pathogen, Human Pathogen, Medical	874	Sanger Institute

Organism	Scientific name	Importance	Genome Size (kb)	Database
Bacteria	*Chlorobium limicola*	Hydrogen production, Biotechnological, Energy production	2434	Joint Genome Institute
Bacteria	*Clavibacter (Corynebacterium) michiganesis sepedonicus*	Agricultural, Plant Pathogen	2941	Sanger Institute
Bacteria	*Erwinia tasmaniensis*	Agricultural	3427	Max Planck Institute
Bacteria	*Escherichia coli*	Human Pathogen, Medical	4126	BCM-HGSC, Univ of Wisconsin-Madison
Bacteria	*Gluconacetobacter diazotrophicus*	Agricultural	3778	LNCC/MCT, UENF, AGROBIOLOGIA, UERJ,UFRJ
Bacteria	*Gluconacetobacter diazotrophicus*	Agricultural	3472	Joint Genome Institute
Bacteria	*Haemophilus somnus*	Medical, Animal Pathogen, Human Pathogen	1980	Joint Genome Institute, Virginia Polytechnic Institute, Univ of Oklahoma
Bacteria	*Mycobacterium marinum*	Medical, Animal Pathogen, Human	5423	Sanger Institute, Univ of Washington, Institute Pasteur
Bacteria	*Cupriavidus taiwanensis*	Agricultural	1031	Genoscope
Bacteria	*Renibacterium salmoninarum*	Fish Pathogen, Medical	3507	Univ of Washington, NCCWA, Integrated Genomics Inc, NWFSC
Bacteria	*Rizobium legumi-nosarum bv trifolii*	Agricultural, Biotechnological	4325	Joint Genome Institute
Bacteria	*Salmonella enterica arizonae sv 62:z4:z23*	Animal Pathogen, Human Pathogen, Medical, Reptile Pathogen	4510	Washington Univ
Bacteria	*Salmonella enterica sv Paratyphi B*	Human Pathogen, Medical	5601	Washington Univ
Bacteria	*Salmonella enterica*	Human Pathogen, Medical	4318	Sanger Institute
Bacteria	*Salmonella enterica Gallinarum*	Human Pathogen, Medical	3965	Sanger Institute, PHLS
Bacteria	*Salmonella enterica sv Paratyphi A*	Human Pathogen, Medical	4078	Sanger Institute, Imperial College
Bacteria	*Candidatus Sulcia muelleri*	Agricultural	227	Univ of Arizona
Bacteria	*Escherichia coli*	Human Pathogen, Medical	4743	J. Craig Venter Institute
Bacteria	*Shigella boydii*	Human Pathogen, Medical	4246	J. Craig Venter Institute
Bacteria	*Thermotoga sp.*	Biotechnological, Evolutionary	1819	Joint Genome Institute
Bacteria	*Ureaplasma parvum*	Human Pathogen, Medical	609	J. Craig Venter Institute
Bacteria	*Ureaplasma urealyticum*	Human Pathogen, Medical	646	J. Craig Venter Institute

Organism	Scientific name	Importance	Genome Size (kb)	Database
Bacteria	*Francisella tularensis*	Medical, Animal Pathogen, Biothreat, Human Pathogen	1406	Los Alamos National Labora tory, Joint Genome Institute
Bacteria	*Xanthomonas campestris*	Agricultural, Plant Pathogen	4471	Bielefeld Univ
Bacteria	*Xylella fastidiosa*	Agricultural, Plant Pathogen	2104	Joint Genome Institute
Bacteria	*Xylella fastidiosa*	Agricultural, Plant Pathogen	2161	Joint Genome Institute
Bacteria	*Yersinia pestis*	Human Pathogen, Medical, Rodent Pathogen	3837	J. Craig Venter Institute
Bacteria	*Anaeromyxobacter sp*	Bioremediation, Biotechnological, Environmental	4457	Joint Genome Institute
Bacteria	*Burkholderia ambifaria*	Human Pathogen, Medical, Agricultural, biocontrol, Biotechnological	7001	Joint Genome Institute
Bacteria	*Burkholderia cenocepacia*	Human Pathogen, Plant Pathogen, Agricultural, Medical	7008	Joint Genome Institute
Bacteria	*Burkholderia multivorans*	Human Pathogen, Medical	6121	Joint Genome Institute
Bacteria	*Burkholderia phymatum*	Agricultural	5421	Joint Genome Institute
Bacteria	*Burkholderia phytofirmans*	Agricultural	7074	Joint Genome Institute
Bacteria	*Methylobacterium populi*	Biotechnological, Agricultural	5314	Joint Genome Institute
Bacteria	*Neisseria meningitidis*	Human Pathogen, Medical	2020	Microbial Genome Center of ChMPH
Bacteria	*Streptococcus pneumoniae*	Human Pathogen, Medical	2155	J. Craig Venter Institute
Bacteria	*Microcystis aeruginosa*	Medical, Environmental, Human Pathogen, Animal Pathogen	6312	Kazusa DNA Research Institute. Univ of Tsukuba
Bacteria	*Rhizobium etli*	Agricultural, Biotechnological	4343	UNAM
Bacteria	*Neisseria gonorrhoeae*	Medical, Human Pathogen	2262	Genotech corp.
Bacteria	*Streptococcus pneunoniae*	Human Pathogen, Medical	2115	J. Craig Venter Institute
Bacteria	*Candidatus Phytoplasma mali*	Plant Pathogen, Agricultural	497	Max Planck Institute
Bacteria	*Leptospira biflexa*	Medical, Animal Pathogen	3667	Monash Univ. Institute Pasteur
Bacteria	*Helicobacter pylory*	Human Pathogen,Medical	1493	Univ of Oregon, Washington Univ
Chicken	*Gallus gallus*		1.5 billions	Sanger Institute, Roslin Institute, Washington Univ
Chimpan-zees	*Pan troglodytes*	Important mammals	48910	BCM-HGSC, Washington Univ, Board Institute

Organism	Scientific name	Importance	Genome Size (kb)	Database
Cryptomonad alga	*Guillardia theta*	Model organism	464	Philipps-Univ Marburg, Canadian Institute of Advanced Research, Univ of British Columbia
Eukaryota	*Candida albicans*	Human Pathogen, Medical	14105	Stanford Univ
Fruit fly	*Drosophila*	*pseudoobscura*	9946	BCM-HGSC
Fungus	*Laccaria bicolor*	Agricultural	20000	Joint Genome Institute
Fungus	*Fusarium graminearum*	Plant Pathogen, Agricultural,	36000	Fungal Genome Initiative (FGI), Board institute
Fungus	*Phytophthora sojae*		95000	Joint Genome Institute, Virginia Polytechnic Institute
Fungus	*Phytophthora romorum*	Agricultural, Plant Pathogen	15743	Joint Genome Institute Virginia Polytechnic Institute
Fungus	*Ustilago maydis*	Agricultural, Fungal Genome Initiative (FGI), Human Pathogen,	6902	BAYER, LION Bioscience AG, Board Institute
Fungus	*Aspergillus niger*	Biotechnological, Citric and production, Fermentation	14086	DSM, Gene Alliance
Fungus	*Aspergillus* (*Emericella*)	Fungal Genome Initiative (FGI), Human Pathogen, Medical, Model organism	9541	Broad Institute
Fungus	*Aspergillus fumigatus*	Human Pathogen, Medical	9923	Sanger Institute, OpGen, Univ of Salamanca, J, Craig Venter Institute, Univ of Manchester, Nagasaki Univ, Institute Pasteur
Fungus	*Candida glabrata*	Medical, Human Pathogen	5397	Genoscope, Institute Pasteur
Fungus	*Neurospora crassa*	Fungal Genome Initiative (FGI)	10097	Univ of Kanas, Univ of Kentucky, OGI School of Science & Engineering, Board Institute
Fungus	*Magnaporthe grisea*	Fungal Genome Initiative (FGI), Plant Pathogen, Agricultural	11109	North Carolina State Univ, Board Institute
Grapes	*Vitis vinifera*	Fruits	475000	INRA, Univ of Mila, Genoscope, Padua
Gray Short-tailed Opossum	*Monodelphis domestica*		19022	Wahington Univ, Broad Institute
Honey bees	*Apis mellifera*	Agricultural, Medical	6704	BCM-HGSC
Human	*Homo sapiens*	Medical	3.0 billion	J. Craig Venter Institute
Mosquitoes	*Encephalitozoon cuniculi*	Medical, Human Pathogen	1996	Univ Blaise Pascal, Genoscope
Mouse	*Mus musculus*	Model organism	39625	Sanger Institute, BCM-HGSC. International Collaboration, Genoscope, Washington Univ

Organism	Scientific name	Importance	Genome Size (kb)	Database
Nematode	*Caenorhobditis elegans*	Model organism	23209	Sanger Institute, Washington Univ
Paramecium	*Paramecium tetrautelia*	Model organism	40000	Genoscope, International Consortium
Plasmodium	*Plasmodium yoelii*	Animal Pathogen, Medical, Rodent Pathogen	7860	NMRC, J, Craig Venter Institute
Plasmodium	*Plasmodium falciparum*	Human Pathogen, Medical	5268	Malaria Genome Project Consortium
Poplar	*Populus trichocarpa*	Agricultural	45000	Swedis Univ of Agricultural Sciences, Joint Genome Institute, Genome Canada, Univ of Washington, Oak Ridge National Lab
Protozoa	*Leishmania infantum*	Medical, Human Pathogen	7993	Sanger Institute, Imperial College, Glasgow Univ
Rat	*Rattus norvegicus*	Model organism	21166	Genome Therapeutics, Univ of Utah, BCM-HGSC, Medical College of Wisconsin, J, Craig Venter Institute, Celera Genomics, CHORI, BCGSC
Rhesus Monkey	*Macaca mulatta*	Mammals	3.0 billion	BCM-HGSC, J. Craig Venter Institute, Genoscope, CUGI
Silk Worm	*Bombyx mori*	Biotechnological, Protein production	18510	SWAU, China
Thale grass	*Arabidopsis thaliana*	Agricultural, Model organisms	26735	International Collaboration
Yeast	*Schizosaccharomyces pombe*	Model organism	5004	Sanger Institute, Cold Spring harbor Laboratory
Yeast	*Sccharomyces cerevisiae*	Model organism	5860	International Collaboration

Source: www.genomesonline.org.

3 Appendix

List of important bioinformatics software and thier web addresses

Application	Software/ Server	Version	Web Address
Database Management	MySQL	5.1	www.mysql.com
	Oracle	11	www.oracle.com
Website Dreamweaver	Adobe	CS 4	www.adobe.com/products/dreamweaver Design
Protein Modeling	Modeller	9v5	www.salilab.org/modeller
	Protinfo		http://protinfo.combio.washington.edu
	SwissModel		http://swissmodel.expasy.org
	PSI-Pred		www.bioinf.cs.ucl.ac.uk/psipred
	THREADER	3	www.bioinf.cs.ucl.ac.uk/threader
	HHPred		www.toolkit.tuebingen.mpg.de/hhpred
Protein Model Validation	Protein Model Check		http://swift.combi.ru.nl/servers/html/modcheck.html
Protein Docking and Molecular Interaction	Autodock	4	http://autodock.scripps.edu
	Hex	5.1	http://www.csd.abdn.ac.uk/hex
	ZDOCK		http://zdock.bu.edu
	ClusPro	1.0 beta	http://nrc.bu.edu/cluster

Application	Software/ Server	Version	Web Address
	Rosetta Dock Server	2.1	http://rosettadock.graylab.jhu.edu
	GRAMM-X	1.2.0	http://vakser. bioinformatics.ku. edu/resources/gramm/grammx
Phylogenetic Analysis	Phylip	3.6	http://evolution.gs.washington.edu/phylip.html
	PAUP	4.0 beta 10	http://paup.csit.fsu.edu
	Mega	4.1	http://www.megasoftware.net
	TreeCon	1.3	http://bioinformatics,psb.ugent. be/software de tails.php?id=3
	MUSCLE	3.6	http://phylogenomics.berkeley.edu/cgibin/muscle input_muscle.py
	DNASIS	2.5	http://www.miraibio.com/products/cat_ bioinformatics/view_dnasismax/indes.html
	DS Gene	1.5	http://www.accelrys.com/dstudio/ds_gene/ index.html
	Winboot		http://www.irri.org/science/softwarewinboot.asp
	NTSYS PC	2.1	http://www.exetersoftware.com/ cat/ntsyspc/ntsyspc.html
	Clustal W	2	www.ebi.ac.uk/clustalw
Sequence Similarity Search	BLAST	2.XX	www.ncbi.nlm.nih.gov
	FASTA		http://www.ebi.ac.uk/Tools/fasta33/index.html
	HMMER	2.3.2	http://hmmpfam.ddbj.nig.ac.jp/top-e.html
	Fasta Search		http://fasta.genome.jp/
Gene Prediction	FGENESH	2	www.softberry.com
	GENSCAN		www.genes.mit.edu/GENSCAN.html
	GLIMMER		http://www.tigr.org/tdb/glilmmerm/glmr_form.html
Molecular Structure Visualization	RASMOL	2.7.4.2	http://rasmol.org/
	Swiss PDB Viewer	4.0.1	www.spdbv.vital-it.ch
	DS visualizer	2	http://accelys.com/products/discovery-studio/ visualization/discovery-studio-visualizer.html
	CN3D	4	www.ncbi.nlm.nih.gov/Structure/CN3D/cn3d.html
	MolMol		http://hugin.ethz.ch/wuthrich/software/molmol/
Molecular Dynamics	VMD	1.8	www.ks.uiuc.edu/Research/vmd

Application	Software/ Server	Version	Web Address
	GROMACS	4	www.gromacs.org
Primer Designe	Primer3	0.4.0	http://frodo.wi.mit.edu/
Sequence Databases	GenBank		www.ncbi.nlm.nih.gov
	EMBL		www.ebi.ac.uk
	DDBJ		www.ddbj.nig.ac.jp
Protein Structure	Protein data Bank		www.rcsb.org
Sequence Assembly	Phred/Phrap/Consed		www.phrap.org
	DNABaser	2	http://www.dnabaser.com/
	DNA Trapper		http://dnptrapper.sourceforge.net
	Sequencher	4.8	www.genecodes.com
QTL Analysis	QTL Cartographer		http://apstc.sun.com.sg/popup.php?ll=researeh&12=projects&13=biobox&f=BioApps&appname=Mapmaker/QTL
	Rqtl	1.09-43	http://www..rqtl.org/
Scripting Language	Perl	5.10.1	www.perl.org
Sequence Alignment	Clustal W	2	www.ebi.ac.uk/clustalw
	SAGA	0.95	http://www.tcoffee.org/Projectshomepage/saga_home_page.html
	MUSCLE	3.6	http://phylogenomics.berkeley.edu/cgi-bin/muscle/inputmulcle.py
Alignment Editor	BioEdit	7.0.9	www.mbio.ncsu.edu/BioEdit/BioEdit.html
	JalView	2.4	http://www.jalview.org/
Motif Search	YMF	3	http://wingless.cs.washington.edu/YMF/YMFWeb/YMFInput.pl
	Gibbs Motif Sampler		http://www.tcoffee.org/Projects_home_page/saga_home_page.html
	BLOCKS		http://blocks.fhcrc.org/blocks/blocks_search.html
Sequence Analysis	EMBOSS	6.0.1	emboss.sourceforge.net
	GCG	11	www.accelrys.com
	DNASTAR	7	http://www.dnastar.com/
	GeneChek		http://www.ocimumbio.com/
SNP Analysis	DNASP	4.50.3	http://www.ub.es/dnasp/

Application	Software/ Server	Version	Web Address
	SNP Detector		http://lpg.nci.nih.gov/
Microarray Data	GeneSpring GX		www.chem.agilent.com/en-US/Products/Software/ lifesciencesinformatics/genespringgx
	Array Assist	1.1.2	http://softwaresolutions.strategene.com/ fdownload/?c=272617
	BioinformatiX		http://www.xpoge.com
Statistical Analysis	MATLAB	8	http://www.mathworks.com/
	R	2.8	www.r-project.org
	SPSS	16.xx	www.spss.com
	SAS		http://www.sas.com/technologies/analytics/ statistics/stat/index.html
Pathways Analysis	Gepasi	3	www.gepasi.org
	Copasi	4.4	www.copasi.org
	Pathway Studio		http://www.ariadnegenomics.com/?id=48
	Pathway Analysis		http://www.ingenuity.com
Chemical Structure Designing	Chemsketch	11	www.acdlabs.com/download

Glossary

Some of the terminologies used in genomics and bioinformatics have been explained below in a simple language. It is not an exhaustive list of the terms. One can find these terms and their descriptions on world wide web.

Adaptation Ability of an organism to acclimatize to the natural environment.

Algorithm A mathematical procedure used to solve complex problems.

Alignment It is a method of aligning two or more protein or DNA sequences to get maximum level of identity and conservation (amino acids). It is used for the assessment of the degree of similarity and the possibility of homology between sequences derived from different organisms.

Anagenesis Evolutionary change along an unbranching lineage i.e. changes without speciation.

Ancestor Any organism, population, or species from which some other organism, population, or species is descended by reproduction.

Ancestral Sequence This is a hypothetical sequence which is reconstructed from the relationships between contemporary sequences. The Maximum Parsimony or Maximum Likelihood can be used for the reconstruction of ancestral sequences.

Assembly Putting sequenced fragments of DNA into their correct chromosomal positions.

BAC Bacterial artificial chromosome, An artificially designed cloning vector which can be used for cloning a medium-sized fragment of a genome (100 to 300 kb) for amplification in the bacteria.

BLAST	**B**asic **L**ocal **A**lignment **S**earch **T**ool.Used for performing sequence comparisons based on database search.
Blocks	These are conserved patterns of about 3-60 amino acids in related proteins sequences.
Bootstrap	It is a method used for the estimation of confidence levels of individuals coming in same cluster during phylogenetic analysis. It uses re-sampling of original data matrix with replacement of characters.
Browser	A programme through which different sites are accessed on world wide web by using a hypertext markup language.
Clade	A monophyletic group formed in a cladogramme.
Cladogramme	A cladogramme is a typical tree which depicts the hypothesized branching order of a number of sequences. Any indication of temporal change cannot be obtained from the cladogramme.
Clustal	The name given to a program used for performing multiple sequence alignment. ClustalW and ClustalX are two important programmes used for multiple sequence analysis.
Cluster analysis	A procedure used for the analysis of a set of related sequences in the form of different groups.
Codon	A group of three adjacent nucleotides of a DNA sequence which encodes an amino acid is known as codon.
Comparative genomics	It means information gained in one organism can have application in the other, even distantly related organisms.
Consensus	A high quality sequence derived from the assembly or multiple sequence alignment of small stretches of similar sequences.
Conservation	Similarity among different types of protein or DNA sequences at different regions as derived from the multiple sequence alignment.
Consistency Index (CI)	CI is a measure which determines how well an individual character fits on a phylogenetic tree. It is calculated by dividing the minimum possible number of steps by the observed number of steps. The average CI value over all of the characters determines the CI values of a tree.
Contigs	Contiguous sequence of DNA created by assembling of the overlapping sequenced fragments of a chromosome.
Cosmid	Another form of cloning vector derived from the bacteriophage and used for cloning large DNA fragments.

Directed sequencing Successively sequencing DNA from adjacent stretches of a chromosome.

Distance Matrix A matrix constructed from the "distance values" obtained from a pairwise sequence alignment.

Diversity The variation in morphology or number of taxa is termed as diversity.

Domain Specific and unique regions of the protein which fold independently and possess its own function.

Domain name A name given to a computer for its classification and identification at different levels of organization of the internet.

Dot Matrix It is a diagram which provides a graphical method for comparing two sequences by placing them on X axis and Y- axis of a graph. The sequence letters which match at specific positions, dots are placed in the graph. The dots present on the diagonal line indicate the alignment.

Draft sequence Sequence with lower accuracy than a finished sequence; some segments are missing or in the wrong order or orientation.

Duplication If a region of DNA is repeated as a copy in other part of the genome then this process is called duplication. If the genes are duplication within a genome, these are called as paralogues.

Dynamic programming It is a mathematical algorithm which breaks complex problems in to smaller one to find optimum solutions.

EST Expressed sequence tag, a unique stretch of DNA within a coding region of a gene. It is useful for identifying full-length genes and also used as a landmark for genome mapping.

Features Annotation added to the different regions of a given sequence.

Firewall A computer which prevents an unauthorized access to the servers of an organization with the help of specific softwares installed in it.

Functional genomics Study of the expression of many genes together as a function of development using high throughput technologies like microarrays.

Gap An indel position within two aligned sequence is called gap. The gaps reintroduced during the alignment so that all homologous positions on the sequences can be aligned.

Gap penalty In a sequence alignment, a numeric score is given to the bases where gaps are created to find optimum alignment.

Genome The entire chromosomal genetic material of an organism.

Genomics — It is defined as the investigations into the structure and function of very large numbers of genes undertaken in a simultaneous fashion.

Global Alignment — The complete alignment of the sequences from start to end is known as global alignment.

GUI — **Graphical user interface.** A graphical softer which allows users to interact with different applications running on a computer.

Heuristic — A simple procedure used to solve complex problems for getting optimum solution.

Hidden Markov Model — A joint statistical model for an ordered sequence of variables is called Hidden Markov Model. Such models are very efficient for bioinformatics analyses because these are based on the algorithm trained for using unaligned or unweighted input sequences for performing analysis of sequence family.

HTML — **Hypertext-markup language.** A text based computer language which specifies the format of document on world wide web.

Hyperlink — A link provided to specific text in a document on world wide web which provide link to the associated documents or pages by clicking on it.

Hypertext — A text which is differentiated from rest of document by giving different colour or by underlining. It functions as a hyperlink between different documents.

Indel — **In**sertion/**del**etion event. During the process of evolution some of the sequences either contracted or expanded at particular points and form gaps while performing sequence alignment.

k-tuple — Short stretches of identical sequences (words) are used for performing sequence alignment.

Leaves — The terminal tips of taxa in a phylogenetic tree are called leaves or termini of the branches.

Likelihood — The probability of occurrence of past events which would yield specific outcome.

Local Alignment — The alignment of small segment of sequences having high similarity compared to the rest of sequences.

LOD score — Log of odds. It is a score which provide statistical estimates to the linked loci on a chromosome.

Machine learning A process of distinguishing alternative probabilities by using computational models is called machine learning.

Maximum Likelihood It is a method of determining phylogenetic relationships among different sequences or individuals.

Maximum Parsimony A method of solving complex problems by using simple solutions. In phylogeny, it means that in a typical phylogenetic tree, a data matrix can be explained by fewer evolutionary events compared to a tree which need more evolutionary events.

Minimum Evolution The function used for optimization of a phylogenetic tree based on a distance matrix. Minimum Evolution can be used to find a tree with the shortest overall branch lengths.

Monophyletic group A group of organisms which shares a common ancestor that is exclusive to these organisms.

Most Parsimonious Tree The fewest number of evolutionary events over entire length of the sequences used for determining branching order in a tree.

Multiple Alignment A multiple sequence alignment is that when more than two sequence are used for alignment.

Neighbour-Joining It is an algorithm used for constructing a phylogenetic tree from a distance matrix by successively clustering pairs of taxa together.

Oligo This term is used for oligonucleotides, which is a short stretch of single stranded DNA.

Orthologue These are homologous genes coding for same function but found in two different taxa.

PAC Phage (P1 from phage) artificial chromosome vector which can contain a genome insert of 100-300 kb size.

Paralogue These are homologous genes coding for same function but found in different parts of the genome of same taxa.

PAUP PAUP is a very important software tool used for the construction of phylogentic tree. It is a standalone programme which can run on different types of operating systems. It can work on any types of binary data derived from the DNA, protein or morphological markers.

PHYLIP A software tool used for the construction of phylogenetic tree from any type of binary data.

Phylogenetics It is a discipline which deals with understanding and resolving relationship among and between organisms.

Phylogram A phylogenetic tree which shows evolutionary relationship between the taxa.

Phylowin A software used for the analysis of phylogenetic relationships among different organisms. Different types of analysis like Parsimony, Likelihood and Distance with bootstrap resampling can be performed with this programme by running on all types of UNIX platforms.

Physical map A map of the locations of identifiable markers spaced along the chromosomes. A physical map may also be a set of overlapping clones.

Plasmid A self replicating circular genetic material present in a bacterial cell. The plasmids can be constructed artificially which is inserted into the bacteria for amplification of DNA to be used in various molecular biology experiments.

Scaffold A series of contigs that are in the right order but are not necessarily connected in one continuous stretch of sequence.

Shotgun sequencing Breaking DNA into many small pieces, sequencing the pieces, and assembling the fragments.

Structural genomics It includes the genetic mapping, physical mapping and sequencing of entire genomes.

STS Sequence tagged site, a unique stretch of DNA whose location is known which serves as a landmark for gene mapping and sequence assembly.

Sum of pairs method A method used in multiple sequence alignment, which is a sum of substitution scores of all possible pair wise combinations of sequence characters in one column.

Taxon Any unit of classification like an individual, a strain or a species is known as taxon. Any types of taxon i.e. the contemporary or hypothetical ancestral strain, species etc. are known as taxon.

Taxonomy A branch of science which deals with the naming of organisms.

Tree length A phylogenetic tree which represents the total number of steps required to map a dataset.

UNIX A computer operating system similar to Windows or MacOS but also has protected memory and scripting features. Most of the genome analysis software works in UNIX environment.

UPGMA Unweighted Pair-Group Method with Arithmetic Averages. One of the first clustering algorithms in which distance matrix is used for dendrogram construction.

YAC Yeast artificial chromosome, yeast DNA which can take in a large fragment of a genome (up to 1 Mb) for amplification in yeast cells.

Index